THE HERCULES

Dedication

To Sir Alfred Hubert Roy Fedden who conceived the engine, to those who developed the engine, to those who maintained the engine, to those who flew the engine to victory in wartime, and so enabled those following to fly the engine in peacetime.

THE HERCULES

THE OTHER ENGINE THAT HELPED WIN THE WAR

GORDON A. A. WILSON

AMBERLEY

Author's Note

All engine facts and statistics in this book should be taken as only being reasonably accurate because even statistics from normally equally reliable sources often differed. Every effort was made to ascertain the correct information from the eighty-year-old material wherever possible.

Half-title page: Sir Roy Fedden by Rick Hovey.

Title page: Hercules sleeve valve engine at IWM Duxford, courtesy of Roland Turner.

First published 2024

Amberley Publishing
The Hill, Stroud
Gloucestershire, GL5 4EP

www.amberley-books.com

Copyright © Gordon A. A. Wilson, 2024

The right of Gordon A. A. Wilson to be identified as the Author of this work has been asserted in accordance with the Copyright, Designs and Patents Act 1988.

ISBN 978 1 3981 1168 4 (hardback)
ISBN 978 1 3981 1169 1 (ebook)

All rights reserved. No part of this book may be reprinted or reproduced or utilised in any form or by any electronic, mechanical or other means, now known or hereafter invented, including photocopying and recording, or in any information storage or retrieval system, without the permission in writing from the Publishers.

British Library Cataloguing in Publication Data.
A catalogue record for this book is available from the British Library.

1 2 3 4 5 6 7 8 9 10

Typesetting by SJmagic DESIGN SERVICES, India.
Printed in the UK.

CONTENTS

Acknowledgements 7
Foreword by Karl Kjarsgaard 10
Introduction 13

1 The Bristol Aeroplane Company 19
2 Sir Roy Fedden 42
3 The Bristol Hercules Pedigree 63
4 Bristol Hercules: The Nuts and Bolts 84
5 The Hercules and Aircraft Companies 116
6 The Hercules Powered Aircraft 136
7 The Hercules in Royal Air Force Service 229
8 The Hercules and The Bomber Command Museum of Canada 252

Conclusion 270
Bibliography 274
Index 281

ACKNOWLEDGEMENTS

My wife Emily for her love and continued encouragement. Without her unconditional support my career in aviation throughout fifty-four years would not have been possible and none of these books would have been completed. My family for their love, encouragement, and understanding when Dad's/father-in-law's/Grandpa's mind seemed to be somewhere else, namely thinking about this book!

Amberley Publishing's management and staff that kept the company going during extremely challenging COVID times, well done. Shaun Barrington, 'editor extraordinaire', thank you for your guidance, patience, and help during the writing process, 'Job Well Done Old Chap'!

Karl Kjarsgaard, H57RC, Bomber Command Museum of Canada, Nanton, Alberta for writing the preface to this book on his favourite engine, being interviewed, and answering many questions, thank you, Karl.

Gary Vincent for editing and preparing the photograph collection for publication and access to his extensive aviation library. Thank you once again Gary, you have been a great help.

The Hercules: The Other Engine That Helped Win the War

Barrie Gabie for his technical assistance, dedicated editing, and grammatical investigations. I am sure that there were long hours with 'hanging participles'! Thank You Barrie.

Syd Leung for his Word support and willingly creating the many tables. ('Not another one! Do you ever write anything?') Thanks Syd.

H.R. (Rick) Hovey for creating the frontispiece drawing. Thanks again Rick.

Colm Egan for many hours researching the world's museums to find examples of the Hercules engine and compiling the list of RAF squadrons that used Hercules-engined aircraft. Thanks Colm.

Chrissie Egan for the beautiful 'Dedication' calligraphy. Nice writing, Chrissie.

Dave Beale for conscientiously researching historic photographs in the public domain, thanks Dave for going through the archives.

Barrie Laycock for aviation research and technical support, thanks Barrie.

Patrick Martin, author, for researching photographic sources.

Bobby Park for information on the Vickers Viking crash site at Irish Law, Largs Moor, Scotland.

Dorothy & Colin Wilson for information on the Hercules engine, National Museum of Flight, East Fortune, Scotland.

Jack Fisher for aviation documentary research.

Jerry Vernon, President, Vancouver Chapter, Canadian Aviation Historical Society.

Nigel Potts for aviation literature research.

Donagh & Eddie O'Brien for continued support of my writing.

Dave Birrell, Chief Librarian & Archivist, Bomber Command Museum of Canada, Nanton, Alberta. Thank you for your help and support.

Canadian Aviation Historical Society, Jerry Vernon.

Aircraft Engine Historical Society.

Avialogs Aviation Library.

Surrey Public Library, Inter-Library Loan department.

Acknowledgements

I would like to especially acknowledge Malcolm Nason and Bill Powderly. Without their 'airspotting' friendship throughout the formative years of my youth, my life's journey in aviation may never have begun. Bicycle, anorak (after all it was Ireland), binoculars, notebook and pencil at the ready.

Images courtesy of Dave Beale, Richard Beale, Doug Bowman, Keith Burton, Colm Egan, Jerry Hughes, Karl Kjarsgaard, David McNeur, John Mounce, Malcolm Nason, Bill Powderly, Roland Turner.

Extensive use was made of three books, all listed in the Bibliography: *British Piston Aero-Engines and their Aircraft* by Alec Lumsden, *Fedden – The Life of Sir Roy Fedden* by Bill Gunston, and *Bristol Hercules Operators' Handbook* by The Bristol Aeroplane Co. Ltd, I am indebted to the authors.

My sincere apologies to anyone I have missed. Rest assured your contributions were much appreciated.

FOREWORD

To write a foreword for this important book on one of the most historic engines in worldwide and Canadian aviation history is an honour. It is important that the story be told of the Bristol Hercules engine, its origins and development, but also of the brilliant people who designed and engineered this power-plant, transforming it into one of the greatest engines in all of aviation and military history.

The great engineer Sir Roy Fedden of England and his perennial design partner Leonard 'Bunny' Butler made up the gifted team that led the way within the Bristol Aero Company for many years before the Second World War, and until victory was assured in 1945. From the very beginnings of aviation and the invention of the motor car in the WW1 era, the motor engineering excellence of Fedden and Butler was second to none.

All the trials and tribulations of engineering, machining. metallurgy, and mass production of auto and aircraft engines were, as the years progressed from 1904–1944, overcome by the Fedden team and Bristol Aero Company to produce great aircraft and engines that would eventually save the free world from the scourge of Nazi tyranny. Not only technical in nature,

Foreword

the political and social roadblocks put in Fedden's way as his career progressed would have discouraged a lesser man, but he persevered in making the best engines for the Allies.

On the Canadian side, the majority of Second World War Royal Canadian Air Force bombers were the Halifax heavy bomber, the greater number of these Royal Canadian Air Force Halifaxes being powered by the Bristol Hercules as part of Canada's contribution to Allied victory.

After my years of research on bomber crews, especially those of the Royal Canadian Air Force flying Halifaxes, I learned that the Hercules engine was what transformed the Halifax into a reliable and tough bomber which could fly 'to-hell-and–back' powered by this Bristol engine. My heroes flew and ran the Hercules engines above all others in the RCAF.

Sir Roy Fedden died in 1973 and my 39-year flying career started in the same year, flying a Bristol Freighter bush plane powered by twin Hercules engines. It is indeed humbling to know I can and will carry the torch for Sir Roy Fedden and his team. Each time we run our rebuilt Bristol Hercules engines at the Bomber Command Museum of Canada in Nanton, Alberta, as a tribute to our Bomber Boys, we are paying tribute also to Sir Roy Fedden and his team.

Enjoy the history contained in this great book, 'The only thing new in this world, is the history you do not know.'

Captain Karl Kjarsgaard
Curator Bomber Command
Museum of Canada
Nanton, Alberta

Author's Note: At the end of a successful flying career with Wardair, Canadian Pacific Airlines, Canadian Airlines, and Air Canada, Kjarsgaard helped start a group called the Halifax Aircraft Association to preserve the history of the Handley Page aeroplane. Upon the successful discovery, recovery, restoration,

The Hercules: The Other Engine That Helped Win the War

and display of a Halifax bomber at the National Air Force Museum in Trenton, Ontario, he looked for another aviation challenge. This time he teamed up with the Bomber Command Museum of Canada, Nanton, Alberta, to restore a Handley Page Halifax bomber powered by Bristol Hercules engines to ground running display condition. Under his enthusiastic leadership, the project is well under way. See Chapter 8.

INTRODUCTION

The silence of the airport was disturbed as the twin Bristol Hercules engines powered the Bristol 170, Mk 31, registration G-AMLJ of BKS Air Transport, into the overcast Autumn skies. It was 1 October 1962 and I stood shivering in the cold rain and wind on the outside viewing area of Collinstown Airport, Dublin. I was a schoolboy aircraft spotter and I was busy collecting registrations of anything that flew. Another registration for the logbook, another successful flight for the Hercules engines.

The momentary silence of another airport and countryside was shattered seven years later at Filton, South Gloucestershire, on 9 April 1969 by four Rolls-Royce/Snecma Olympus engines in reheat. It was the first flight of the number two Aerospatiale/British Aircraft Corporation's supersonic passenger airliner Concorde. Fifty-eight years previously, in 1911, George White, later Sir George, of British Tramways, had established a 'flying ground' opposite Fairlawn Avenue at the top of Filton Hill for the use of the British & Colonial Aeroplane Company. The grass strip became Filton airport. The connection? Filton was the birthplace of the Bristol Hercules engine.

The British & Colonial Aeroplane Company evolved into the Bristol Aeroplane Company (BAC) and initially set out in early

The Hercules: The Other Engine That Helped Win the War

1910 to build aeroplanes. After a few false starts it produced the successful Bristol Boxkite in July 1910. This aeroplane was followed by the Bristol Scout and Bristol F.2B Fighter during the First World War. Its reputation established, BAC had the opportunity to invest in the aero-engine division of Cosmos Engineering Company in 1919. The Company now had a commanding position in the post-war aviation scene, as both an aircraft manufacturer and a supplier of aero-engines. BAC inherited the talents and designs of Roy Fedden. One of these designs being the very successful air-cooled, nine-cylinder, radial Jupiter engine.

Fedden, later Sir Roy, had disappointed his wealthy family's expectations by deciding on an engineering career. Initially engineering cars, he progressed to the repair of aero-engines. He would continue to develop aero-engines in the inter-war years based on his pre-war experience with Cosmos Engineering; first with the normal poppet-valve operated engines, the Bristol Mercury and Pegasus engines, and later with the revolutionary sleeve-valve engine. Fedden had read papers by Harry Ricardo of the Royal Aeronautical Society in the 1920s extolling the virtues of the sleeve-valve engine. It had caught his attention, and he turned his creative engineering genius to investigating this revolutionary approach to engine design.

Three names are associated with the invention and refinement of the sleeve-valve engine, Charles Knight in the USA (1904), and Peter Burt and Harry McCollum in Great Britain (1909). Burt and McCollum were employed by the Argyll motor car company. It is interesting to note that Bristol, Napier, and Rolls-Royce produced sleeve-valve engines. Bristol designed the single-row, sleeve-valve Aquila, and Perseus engines. These engines had a short useful life span due to the demand for more powerful engines as designers required aeroplanes to grow bigger and fly faster. Fedden's answer was the two-row, sleeve-valve, Bristol Taurus, and Bristol Hercules engines. The aero-engine world was dividing into air-cooled, water-cooled, poppet-valve, and sleeve-valve designs – not to mention radial, opposed, Vee, and inline cylinder configurations.

Introduction

What made the Bristol Hercules so different from the other engines of that time? The other peer engines used a system of a camshaft, pushrods, rocker arms, and valves embedded in the cylinder head that allowed the fuel/air mixture to enter, be compressed, be combusted for the power stroke, and be exhausted. The Bristol Hercules used an ingenious single-sleeve valve that moved vertically and had partial rotary motion in the cylinder bore to achieve the same four engine cycles. The sleeve had ports (gaps) cut in it, which functioned as valves as the sleeve's vertical and rotary motion covered and uncovered its ports in sequence.

The Perseus engine evolved into the Hercules, which first ran in 1936 and was available in 1939 as the 1,290hp (960kW) Bristol Hercules I engine. Initially, the engines were hand-built so that the tolerances would be acceptable, but this was time-consuming and not suitable for mass production. The main challenge was to maintain sufficient cylinder and sleeve lubrication. Just before the start of the Second World War a method of centrifugal casting was developed to make the sleeves perfectly round, to maintain tolerances and thus allow mass production. The advantage of the sleeve-valve engine was that it allowed a higher compression ratio for specific octane levels resulting in greater power output. Bristol developed a modular 'unitised' engine installation for the Bristol Hercules allowing the engine and cowling to be installed in any suitable aeroplane.

The Royal Air Force, the world's oldest independent air force, was formed in April 1918 on the illustrious foundations of the First World War Royal Flying Corps. In the late 1930s it was expanding rapidly due to the ever-increasing threat of war from Nazi Germany. In 1936 the RAF established Fighter, Bomber, and Coastal Commands. These Commands would be followed in 1943 by Transport Command. By 1939, the need for aeroplanes and engines now became critical. Bristol with its Bristol Hercules engine and Rolls-Royce with its Merlin engine would provide a large share of the piston wartime requirements, and for a time post-war, until superseded by the jet engine. The Bristol Hercules powered four RAF bomber aeroplanes and one fighter.

The Hercules: The Other Engine That Helped Win the War

The Bristol Hercules engine was installed in over thirty aircraft, some of which were very successful – the Handley Page Halifax for example – and others not so. For instance, the Armstrong Whitworth Albemarle was designed in response to Specification B.9/38 for a medium bomber. However, its role was altered to that of aerial reconnaissance and transport duties midway through construction. This delayed its entry into service until 1943, to the point that it was no longer credible to be considered as a bomber. Other engines were installed as a one-off for engine testing, such as the Northrop Model A8-1, or for trial purposes in the Avro York C.II.

A.V. Roe and Company (Avro) used the Bristol Hercules in very limited numbers in their operational Avro Lancaster heavy bomber, the B.II model only, and converted one Avro York from the Merlin engine to the Bristol Hercules for test purposes. The Company used the Rolls-Royce Merlin as the powerplant of choice for their Lancaster bomber, installing over 7,000 engines. There had been a concern over shortage of Merlins and the limited installation of the Bristol Hercules engine possibly established the feasibility of it as a backup powerplant for the Lancaster.

The Bristol Aeroplane Company was one of the few companies that not only designed and built aircraft but designed and built the engines to power them. The engine division was the result of the Cosmos Engineering takeover in 1920, Roy Fedden was an employee of Cosmos. The rest as they say, is history. The famous single-seat biplane fighter Bristol Bulldog (1929), for example, used its own Bristol Jupiter engine. The Bristol Hercules was no exception. It saw wartime service in the multi-role Bristol Beaufighter and post-war in the Bristol 170 freighter and airliner series. The concepts of the Jupiter, piston size, and Bristol Hercules engines would be further developed into one of the most powerful wartime piston engines, the Bristol Centaurus.

Handley Page Limited was an early – 1909 – aeroplane manufacturing company in Great Britain. It had previously used other manufacturer's engines for its aeroplanes and for the

Introduction

Handley Page Halifax it used both the Bristol Hercules and the Rolls-Royce Merlin engines. The Bristol Hercules was used in the H.P.61 Halifax B.III, the main production variant of the heavy bomber. The tail-wheel H.P.67 Hastings military transport and the tricycle H.P.81 Hermes civilian airliner both used the Bristol Hercules; the engine was showing its versatility. The Hermes mainly used the 763 variant and the Hastings mainly used the 101 and 106 variants.

Short Brothers plc was the first company in the world to make production aeroplanes. In the 1920 and 1930s it successfully diversified into flying boats. The 1939 S.26 'G' class flying boat designed for non-stop trans-Atlantic flights, the 1944 S.45 Seaford long-range maritime patrol bomber for Coastal Command, and the 1946 S.45 Solent passenger flying boat all had Bristol Hercules engines. The 1939 S.29 Stirling was the first four-engine, heavy-bomber for the Royal Air Force and used the Hercules. The engine had established its reliability for military and civilian long-range operation.

Vickers (Aviation) Ltd used the Bristol Hercules engine in five of their aeroplanes, one Second World War four-engine bomber – the Wellington – one 1930s engine test bed – the Wellesley – and the post-war RAF Valetta transport, RAF crew-trainer Varsity, and Viking VC.1 civil airliner. The Wellington bomber also used the comparable Rolls-Royce Merlin and Pratt & Whitney Twin Wasp engines.

Other aeroplanes powered by the Bristol Hercules were one-off engine test bed variants, one Spanish transport, one Saunders-Roe flying boat, and two French post-war transports. One of the transports, the Nord Noratlas, was the last flying example of the Bristol Hercules engine in the world, seen during the French 2019 airshow season. Hopefully, it will return to the skies in 2022.

There are a few museum exhibits and ground stand-based running examples of the Bristol Hercules extant. The topic of one of the chapters here concerns the restoration of four Bristol Hercules engines to be installed at a later date and ground run in a restored, displayed Handley Page Halifax at the Bomber

The Hercules: The Other Engine That Helped Win the War

Command Museum of Canada in Nanton, Alberta. The Museum currently has a ground, stand-running, Bristol Hercules for public events.

This book will take the reader through the story of the aero-engineer who developed the Bristol Hercules, Roy Fedden, the company that backed him, Bristol Aeroplane Company, and provide a layman's description of the engine, the engine's contribution to the Allied victory in the Second World War, and its continued contribution to the post-war success of civil and military aviation. The Bristol Hercules is your story, Sir Roy. As they politely say in England, 'Well done, old chap': Sir Alfred Hubert Roy Fedden, MBE, FRAeS (6 June 1885-21 November 1973).

I

THE BRISTOL AEROPLANE COMPANY

Bristol, an ancient city, is situated on the River Avon, which flows into the Severn Estuary connected to the Bristol Channel, giving Bristol access to the oceans of the world. The city quickly established itself as a major maritime centre and during the fifteenth century was the second most important port in England, trading with Ireland, Gascony (France), and Iceland. Voyages of exploration departed from Bristol to explore the New World and John Cabot made landfall in North America from there in 1497. The ships were getting bigger, so the River Avon was no longer accessible and the Avonmouth Docklands were established.

Bristol by the turn of the twentieth century had established itself in the world of commerce with several families seizing the available commercial opportunities. Bristol was the name given to one of the most important British aero-engine and aircraft manufacturers of the twentieth century, founded by George White. This company would succeed and contribute to both the First and Second World Wars. It was fitting that an aviation company would follow in the footsteps of a long and successful maritime tradition based in Bristol. The aeroplanes, too, would 'sail' the world, as their maritime predecessors did.

The Hercules: The Other Engine That Helped Win the War

A growing city needed communication and transportation to function efficiently, and George White saw an investment opportunity. White was a businessman and a stockbroker who first became involved in transportation with his work at the Bristol Tramways Company. Initially the trams were horse-drawn, but by 1895 the system had become electrified, the first city in Britain to do so. At its peak Bristol Tramways had seventeen routes and 237 tramcars. Operations ceased in April 1941 due to major damage by a Luftwaffe air raid, the 'Bristol Blitz'. Not only was White proving to be a shrewd investor, but he was also rubbing shoulders with the financial elite of Bristol, which gave him access to funds for future ventures. One of these was the W. D. & H. O. Wills tobacco company. Their first brand of cigarette was called 'Bristol', what else, made at the London factory from 1871 to 1974.

White invested in and owned tram companies in other cities and had interests in railway connections in the area. The tram investments led to investments in transportation industry equipment such as new double-decker buses, vans, lorries and commercial vehicles. In 1904, White became Sir George White, 1st Baronet. This was a few short months after two American brothers proved that air transportation was feasible – and that future development might make it practical. Who were the industry leaders to do that? Sir George White would be one of them. Six years would pass before White focussed on the aviation industry.

In 1909 while on a visit to France, he witnessed Wilbur Wright demonstrating the Wright Flyer. The astute investor and businessman saw the potential of this new industry and upon returning to England formed four companies. He immediately introduced the words Bristol, British, Colonial, aviation, and aeroplane in his new company names. 'All bases were covered' upon his entry into the emerging aviation industry. Bristol, Filton and the surrounding area would never be the same. The many years of seafaring tradition would now be followed by a newcomer, the aviation industry.

White formed four companies on 19 February 1909. They were The Bristol Aeroplane Company Limited, The Bristol Aviation

The Bristol Aeroplane Company

Company Limited and The British and Colonial Aeroplane Company Limited, all funded mostly by the White family, and The British and Colonial Aviation Company Limited. The only financial option at the time for start-up was private funds as the Bristol Stock Exchange was not initially supportive in this new venture. Things would change with the advent of the First World War.

An inter-White company leasing agreement resulted in the Bristol Tramways Company's repair shed being used by the fledgling aviation company. The new premises were at the end of the tramline in Filton, 7 kms (4 miles) north of Bristol's city centre. Where would the company go to get started in the aviation industry? France, at that time, was the leader in European aviation having continued its historic development from the balloons of the Montgolfier Brothers. Many aviation terms have their origin in French, such as pitot and aileron. White looked to the French aviation industry and a manufacturing licencing agreement was signed with Société Zodiac of Paris to build their biplane aeroplane, facilitated by their agent in France, Emile Stern. An example of the Zodiac was brought to England, exhibited at the Olympia Aero Exhibition in March 1910, and sent to Brooklands, Surrey, for further evaluation. Brooklands was at the centre of automobile testing, its banked racetrack specifically built for British automobile manufacturers. The large, flat centre of the racetrack was an ideal place for the aviation pioneers, including A. V. Roe and Tommy Sopwith, to gather and test their products.

The Zodiac was a product of the Voisin brothers, former manufacturers of balloons and small dirigible airships. In 1907 they had built a 'boxkite' type of biplane and the next year won a competition to fly around a one-kilometre closed course. Not only had the biplane stayed in the air a longer distance, but it had successfully made three turns. The pilot was Henri Farman. Farman was an adventurer who had started off in cycling races, progressed to automobile racing, and finally turned his hand to piloting aeroplanes.

The 1907 Voisin biplane was a success, the Zodiac was not! Unfortunately, the Zodiac was a complete failure for the

The Hercules: The Other Engine That Helped Win the War

company, it could not get it to fly at Brooklands even after further 'flight tuning'. The licencing agreement and five Zodiacs under construction were cancelled and, upon the recommendation of the pilot testing the Zodiac, the company's attention turned to the French Henri Farman aeroplane. The Zodiac had paid attention to detail and did not fly, the Farman was of rougher construction and did fly.

The British and Colonial Aeroplane Company Limited would try again to get airborne. Under the newly appointed Engineer and Works Manager George H. Challenger's guidance, the company took the basic Farman design, improved its construction and finish, and by July 1910 had their own version in the skies piloted by Maurice Edmond. Initially, Farman Freres were going to sue for patent infringements but dropped the case as they realised that the Bristol-based company had substantially improved the design and the case would be difficult to prove in court. The Bristol Biplane became known as the familiar, but inaccurate, Bristol Boxkite. The British and Colonial Aeroplane Company Limited (the Company) was now truly in the aeroplane manufacturing business.

The Company had leased 924 ha (2,284 acres) of land at Larkhill, Wiltshire on Salisbury Plain from the War Office and tested the new aircraft there. Simultaneously, it concluded that there would be a requirement to teach customers to fly their products, so it established flying schools in 1910 at Larkhill and Brooklands with the associated sheds and hangars. By 1914, out of over 600 certificates issued by the Royal Aero Club, 308 had been issued by the two flying schools. The nearby army camps could not help but become interested in the activity overhead in the skies at Larkhill: a subtle and eventually effective way of advertising to the War Office. By November 1910, two Boxkites a week were being built at Filton. In December 1910, the Company donated a Challenger-designed Bristol Glider to the Aero Club.

An overseas order was received from Russia, and sales promotion events were held in India and Australia. The War Office set aside their objection to being involved in aerial warfare and submitted an order for four Boxkites in March 1911. Even the government realised it was getting left behind in aviation developments. These aeroplanes, the first two with 37 and 38 construction numbers, would be the first British aircraft for the Army Air Battalion. By January 1911, the Filton factory had nearly 100 employees. In just over two years the Company was producing and selling its own aircraft abroad and at home. Quite an achievement for a family-run tram company. Bristol was now on the aviation world map.

During the next ten years the Company would build a variety of aircraft, some examples below, using what engines were available from engine manufacturers. Initially, the only aero-engines available were European. ENV was Anglo-European, but by 1916 British engines began to be installed in Company aeroplanes. The Bentley, named after the automobile engine designer, AR (Admiralty Rotary) 1 was one of the first British engines.

Bristol Aeroplanes 1910–1919

Year	Aeroplane	Engine
1910	Bristol Biplane/Boxkite	37 kw (50 hp) Gregoire, 37/45 kw (50/60 hp) ENV, 37/52 kw (50/70 hp) Gnome, 45 kw (60 hp) Renault
1911	Bristol Monoplane	37 kw (50 hp) Gnome
1911	Bristol Biplane Type T	45 kw (60 hp) Renault, 52/75 kw (70/100 hp) Gnome
1912	Bristol Side by Side Monoplane	37/60 kw (50/80 hp) Gnome, 52 kw (70 hp) Daimler
1913	Bristol-Coanda Two-Seat Biplane	37/60 kw (50/80 hp) Gnome, 52 kw (70 hp) Renault, 45/60 kw (60/80 hp) Le Rhone, 56/75 kw (75/100 hp) Mono-Gnome

The Hercules: The Other Engine That Helped Win the War

Year	Aeroplane	Engine
1914	Bristol Scout Types A, B, C, D, SSA, S2a	60 kw (80 hp) Gnome, Le Rhone, Gnome Lambda, Clerget 75 kw (100 hp) Mono-Gnome 82 kw (110 hp) Clerget, Le Rhone
1916	Bristol M1 Monoplane Scout	82/97 kw (110/130 hp) Le Clerget 97 kw (110 hp) Le Rhone 112 kw (150 hp) Bentley AR1
1916	Bristol F2A, F2B	164/205 kw (220/275 hp) RR Falcon 112/149 kw (150/200 hp) Hispano-Suiza 149 kw (200 hp) Sunbeam Arab 149 kw (200 hp) RAF 4d 134 kw (180 hp) Wolseley Viper 171 kw (230 hp) Siddeley Puma 224 kw (300 hp) Hispano-Suiza
1917	Bristol M11 Metal Biplane	104 kw (140 hp) Hispano-Suiza 134 kw (180 hp) Wolseley Viper
1918	Bristol Scout E, F	149 kw (200 hp) Sunbeam Arab 235 kw (315 hp) Cosmos Mercury
1918	Bristol Braemar, Pullman, Tramp	172 kw (230 hp) Siddeley Puma 298 kw (400 hp) Liberty 12
1919	Bristol F2C Badger I, II, X	239 kw (320 hp) ABC Dragonfly 298 kw (400 hp) Cosmos Jupiter 172 kw (230 hp) Siddeley Puma
1919	Bristol Tourer	205 kw (275 hp) RR Falcon III 172 kw (230 hp) Siddeley Puma
1919	Bristol Babe	34 kw (45 hp) Viale 30 kw (40 hp) Siddeley Ounce 45 kw (60 hp) Le Rhone

The business decision now was whether the Company would continue to use other manufacturers' engines or take the huge leap of faith and design their own engine and create a 'Bristol package'. Aero engines had increased in power from 80 hp to 400 hp during the First World War, the potential was there for an investment of cash and talent, but who would take the Bristol company down this aerial path to the future? The Company realised that it would have to develop lighter and faster products to keep up with the competition.

George Challenger and Collyns Pizey, his assistant, went to Larkhill to learn how to fly before establishing an Experimental Department. Pizey stayed on at Larkhill as manager and Challenger teamed up with Archibald Low at Filton. Low was former manager at Brooklands. Gabriel Voisin, of Zodiac fame (infamy), was hired as a consultant! By the end of 1911 the Bristol-Prier Monoplane was being used for advanced instruction at the Bristol flying schools. Note that this aeroplane was a monoplane. Perhaps a revolutionary step too far by the construction and aerodynamic standards of the day? Unfortunately, the Company was also losing technical staff poached by other aviation companies and had to continue to hire new employees.

It was necessary to produce a biplane design, as the War Office had banned the Royal Flying Corps from flying monoplanes effective September 1912, due to increasing accidents and fatalities. This precluded any military orders for the monoplane in Britain. In 1913, the Company teamed up with a Romanian engineer, Henri Coanda, to produce a biplane. Coanda had already designed and produced a monoplane.

The two Company flying schools at Filton and Brooklands continued to flourish as their reputation and fame spread; not only did countries model their initial flying schools after them, but the countries themselves became interested in the Bristol aeroplanes. Commercially, the Company was still progressing at the beginning of the First World War.

Frank Barnwell, a marine engineer, and his assistant Clifford Tinson were working on an aeroplane that could operate on water. Valuable information was gained with the prototypes, but the project stalled. Barnwell, now gaining experience as an aeronautical engineer, collaborated with pilot Harry Busteed and produced the classic Bristol Scout. By the end of 1914, the War

The Hercules: The Other Engine That Helped Win the War

Office started to order the Scout variants and more than 370 were built to serve with the Royal Flying Corps and Royal Naval Air Service. It was built at another old Bristol Tramway works at Brislington, another suburb of Bristol.

At the same time there was a feeling among the young male employees that 'they were not doing their bit' for the war effort and they left the Company voluntarily to join the military, including Barnwell. This was a major loss to the Company. However, by the end of summer 1915, Barnwell was back at Filton on indefinite leave in his old job as Chief Designer. It was decided that he was more valuable to the war effort as the Company's aeroplane designer than as a 12 Squadron, Royal Flying Corps pilot.

By August 1916, the factory was deemed a 'controlled establishment' by The Munitions of War Act 1915, at the same time as conscription was introduced. This was a British Act of Parliament passed on 2 July 1915. It was designed to maximize munitions output and brought private companies supplying the armed forces under the tight control of the newly created Ministry of Munitions. There would be no labour disputes, and the factory was given maximum support by law to produce its product – in the Company's case, aeroplanes. In effect, the factories were temporarily nationalised.

It was a British engine, the Rolls-Royce Falcon, combined with Barnwell's leadership and design team, that created the successful Bristol F2A and F2B Fighter. The aeroplane first flew in early 1916, an event overshadowed by the death of the founder Sir George White in November. His brother Samuel White took over. In July 1917, finally, the War Office recognised the achievements of Bristol in the war effort and it had the confidence to equip all fighter-reconnaissance squadrons with Bristol Fighters. More than 5,000 F2 aeroplanes were built over its lifetime.

This necessitated the use of sub-contracts for parts from the car industry and engineering companies. The airframes produced soon outstripped the available Rolls-Royce Falcon engine production, so that Sunbeam, Hispano-Suiza, Wolseley and

Siddeley engines were substituted, not always successfully. The aeroplane was built under licence after 1917 in the US, with the Wright-Hispano-Suiza version being the most successful version.

In the last year of the war, other products were investigated and produced, such as the Porte flying boat, the single-seat Scout, the Braemar bomber, and the Badger. The Armistice on 11 November 1918 changed everything. For humanity it was happy news, for the Company it was a challenge. Vast surplus stocks of aircraft and parts were available as demand disappeared overnight, leaving just a few contracts to be completed by 1919.

The Company decided to remain in the aviation business and looked to futuristic designs to keep them competitive. Barnwell was looking for an engine to replace the Sunbeam Arab in the Scout Type F. The Cosmos Mercury engine was installed, and it flew in September 1918 and was tested at the Royal Aircraft Establishment. Another Cosmos engine, the Jupiter, in May 1919 flew in a prototype Bristol Badger. In June 1919, the Company built a wind-tunnel; Barnwell insisted that it was necessary to continue project development.

The Company's challenge changed overnight from supplying war material to trying to stay in business, with no immediate demand for its existing products. Forays into coach building for cars, buses, light car production, and a small motorcycle engine were attempted. The idea was to keep the nucleus of the skilled workers intact for future developments. Fortunately, as it turned out, Barnwell had maintained professional contact with another Bristol aviation company, Cosmos Engineering (Cosmos) of Fishponds, Bristol. Roy Fedden was chief engineer for Cosmos.

Fedden had previously designed for Cosmos three successful air-cooled radials, the Mercury, Jupiter, and Lucifer. In February 1920, Cosmos went into liquidation (see below, page 49). This was not because of their aviation interest but a shipping deal that went wrong. The Air Ministry strongly suggested to Bristol that it would be in their best interest to take over Cosmos, which it did, and subsequently made it the nucleus of their Aero-Engine Department. The takeover included some completed engines and tooling.

The Hercules: The Other Engine That Helped Win the War

Bristol now had the 'all British' package – an aeroplane and engine manufacturer producing the total product. The Jupiter and Lucifer engines were exhibited on the Company's stand at the August 1920 Aero Show at Olympia, London. This was the turning point for the Company's post-war recovery and future aviation industry success; time for a new identity. The British and Colonial Aeroplane Company had existed for ten years and was unable to register the name 'Bristol' as a trademark for aircraft products. The decision was made to voluntarily liquidate the company after its massive, successful growth, and to form the new Bristol Aeroplane Company Limited (Bristol) which had the Bristol name. The name would be synonymous with the best of the British aviation industry in the coming years, including the yet unimaginable Second World War.

Who was this Roy Fedden of Cosmos and why would he have such an influence on the future of Bristol? Chapter 2 tells his personal story of success, his contribution to Bristol, and provides some personal anecdotes from behind the scenes. In this chapter we will continue with the story of Bristol, including the influence of the Fedden partnership.

At the end of the war, Bristol had 3,000 employees in its production works at Filton and Brislington. A formidable company, but Government wartime money and contracts had dried up as the Services re-examined their requirements in peacetime. As far as Barnwell was concerned, the immediate way forward was a single-engine, metal monoplane aeroplane that was role-convertible. It was the right idea, but the timing was wrong. Too many war surplus biplanes were available and there was no government money for monoplane prototypes. A disheartened Barnwell left to take employment with the Royal Australian Air Force in October 1921. By 1923, Barnwell had returned disillusioned with Australia and resumed his post as Chief Designer.

It would be several years before Bristol's aero-engine division showed any profit, but it would be the Jupiter engine that led the company out of the business doldrums. Indeed, Bristol

The Bristol Aeroplane Company

would eventually dominate the air-cooled radial engine market during the inter-war years. As well as Bristol supplying its own aeroplanes, the Jupiter and its successors would power many aeroplanes built by other manufacturers. For a list of manufacturers using Bristol engines see Chapter 3.

It would take until the mid-1930s, and the looming threat of war, before Bristol could adopt the new metal monoplane structure that Barnwell had been advocating and which finally made the biplane obsolete. Wilfred Reid had assumed the position of Chief Designer in Barnwell's absence and, with Fedden, got the new Jupiter engine flying in a ten-seat commercial transport aeroplane. Good news financially for the company was that the Jupiter engine was also manufactured under licence by Gnome and Le Rhone. Additional revenue came from the servicing of existing aeroplanes and engines for the RAF.

But what of the successes, and failures, of Bristol to continue to produce competitive aeroplanes in a wide-open market with no government contracts? We will highlight some of these aeroplanes as they contributed, or did not, to the overall financial health of the company. From the end of the First World War to the end of the Second, the company would produce over thirty aeroplanes. Some would only be prototypes and demonstrators, some would be extremely successful in peacetime, some would initially be private order (non-government), while others would contribute immensely to the success of the WW2 effort.

For example, Bristol, as a private order, built a twin-engine transport aircraft that was eighty kilometres per hour (fifty miles per hour) faster than the RAF's fighter aircraft! This type of aeroplane had cantilever wings, metal monocoque fuselage, retractable landing gear, powered by supercharged engines driving variable-pitch propellors. This, the Type 142, would evolve into the very successful Blenheim bomber.

The following aeroplanes are mostly examples of biplanes built by Bristol during 1920 until 1934, when finally production would switch totally to the monoplane design. The private venture Bulldog in 1927 was by far the most successful aeroplane produced. It was the mainstay of the RAF fighter force from 1930 to 1937 – unbelievable that a biplane would be still part of the inventory with the Second World War only two years away! The Bulldog did not have any selling restrictions to other countries as it was a private venture by the company. It was part of the fighter force for Denmark, Estonia, Finland, and Australia, and was a commercial success.

1920–1934

Bristol Prototypes/Demonstrators

Bullet (1920) was designed with two purposes, to test the Cosmos Jupiter engine under high-speed manoeuvre conditions and publicise the Company in international races. It was one of the fastest aeroplanes at the time, 258 kph (160 mph). It was exhibited in December 1919 at the Paris Salon with a mock-up engine. Production: 1

Bullfinch (1922) was built as a monoplane fighter and a biplane reconnaissance, gunner's cockpit, aeroplane experimentally for the RAF. Production: 2 fighters, 1 Biplane

M.1D (1922) had the Bristol Lucifer engine and the monoplane was entered in various Aerial Derbies. In the June 1923 Grosvenor Cup Race it crashed, suspected bracing wire failure, killing the well-respected pilot, Leslie Foot. Production: 1

Racer (1922) was designed by Fedden and Reid after Barnwell's departure for Australia. The monoplane had the potential of 322 kph (200 mph) but was plagued with wing torsion, lateral instability, aileron control, and major spinner problems. Production: 1

Brownie (1924) was a light tandem-seat monoplane using the Cherub engine. It was well engineered, efficient in performance,

The Bristol Aeroplane Company

economical in fuel usage, but was not rugged enough for flying club usage. Production: 3

Type 92 (1925) was a laboratory aeroplane which tested every part of its construction by experimenting with metal, wood, fairings, and every innovative design of its time. Production: 1

Badminton (1926) during a pre-King's Cup Race flight the engine seized and the aeroplane crashed killing Captain Frank Barnard of Imperial Airways. Production: 1

Type 101 (1927) came in second place at the 1928 King's Cup Race. It replaced the Badminton as a test aeroplane. It was the last wooden aeroplane produced at Filton. It crashed during an overspeed test; the pilot survived. Production: 1

Bristol Bombers
Berkeley (1925) was built with a Rolls-Royce Condor III engine but succumbed to the Air Council decision to not use any single-engine night-bombers.

Bristol Fighters
Bloodhound (1923) was instrumental in the Jupiter versus Jaguar engine competition. The Jupiter was flown for 225 hours between Filton and Croydon to demonstrate the reliability of the engine to Imperial Airways. It also did a Croydon-Cairo flight return, 8.916 km (5,540 miles) in fifty-six hours. The only delay being desert landing grounds, not engine-related. Production: 4

Jupiter-Fighter/Trainer (1923) was derived from the F2B airframe, installing a Jupiter engine. Successfully displayed at the Gothenburg International Aero Exhibition, winning arrival, climb, and aerobatic prizes. It also demonstrated its cold-weather capability in Lapland with resulting sales to the Swedish government. It was used in flying schools with success using various Marks of the Jupiter engine. Production: Fighter 3, Trainer 23.

Boarhound/Beaver (1925) was no match for the Armstrong Whitworth Atlas and with continuing modification ended up serving in Mexico. Production: 4

Bagshot (1927) was a night-fighter that suffered from aileron control reversal caused by wing torsional flexibility at higher speeds. Production: 1

Bulldog (1927) was the most successful aeroplane for the RAF during the inter-war period. The unequal span, single bay biplane was powered by many Bristol engines including the Jupiter, Mercury, Perseus, Aquila, and some power plants from other manufacturers. It was of all-metal construction covered in fabric and had two Vickers guns mounted each side of the cockpit. The Bulldog had a short-wave wireless transmitter-receiver, and durability and ease of maintenance made it a winner. Bristol received orders from Australia, Denmark, Estonia, Finland, Japan, Latvia, Siam, and Sweden which helped the Company hugely financially. Bulldogs saw combat with the Finnish Air Force in the Winter War against the Union of Soviet Socialist Republics. It also served overseas with the RAF in Sudan during the Abyssinian crisis. Production: 443

Bullpup (1928) was used as an engine test bed. Production: 1

Type 123 (1934) was the last Bristol biplane built at Filton. The specification was for better high-level performance, exceptional manoeuvrability, longer endurance, good visibility, low landing speed, four synchronised Vickers guns, two-way radio, and to use the Rolls-Royce Goshawk engine! It had high speed lateral instability and the design was abandoned.

Bristol Commercial Transports
Seely (1920) entered in an Air Ministry competition to encourage designs for safety and comfort. Points were awarded for payload and economy, ability to land slowly, and takeoff with a minimum

distance. Converted into a laboratory aeroplane with a Jupiter III engine it later saw service with the Royal Aircraft Establishment. Passengers: 2, Production: 1

Ten Seater/Brandon (1921) modified with a Jupiter IV engine it saw service with Instone Air Line and later as a freighter on the London-Cologne route. The Brandon was a troop carrier and air ambulance for the RAF. Passengers: 8, Production: Ten Seater 2, Brandon: 1

Taxiplane/Trainer (1923) had two side-by-side passenger seats behind pilot and the Trainer had two seats in tandem. Passengers: 2, Production: 3 Taxiplane, 24 Trainer (Lucifer), 1 (Titan)

Type 109 (1929) was designed to give Bristol publicity with its Jupiter engine in an attempt at the world's long-distance record. Pilots: 2, Production: 1

Type 110A (1929) used the Titan engine. Passengers: four, Production: One completed

Bristol General Purpose
Type 118/120 (1931) was Barnwell's private venture to provide a high-performance general purpose (fighter, bomber, recce), two-seater to smaller air forces. Seats: two, Production: 118 1, 120 1

In 1934, the obsolete Bristol biplanes gave way to the sleek, stressed-skin monoplane Type 133. The government (more or less) abandoned political appeasement in 1935 and embarked on a program to modernise the RAF. The German threat could not be ignored any longer. Filton now entered another expansion mode. The number of employees doubled to 4,200, the erecting shop doubled in area, tool and machine shops expanded, a new paint spray shop was added, and a storage area for the new Alclad sheets was constructed. Next to Filton House, a new head office block was built to oversee the whole manufacturing process.

The Hercules: The Other Engine That Helped Win the War

A new factory specifically for the engine department was built east of Gloucester Road. It covered 18,581 sq m (200,000 sq ft) and allowed for further additions if necessary. The Rodney Works was exclusively for engine and cowling manufacturing. The Bristol Company now covered 296 ha (732 acres) and made Bristol, in just over twenty-five years, the world's largest aeroplane manufacturing factory.

The RAF's requirements now far exceeded the capacity of the aviation industry, even after its expansion at the request of the politicians. Who or what could provide the required extra capacity? The obvious answer was to turn to the industry which had provided the initial engineers and works, the automotive industry. The 'shadow scheme' was born, with Bristol Mercury engines being made at Patchway near Bristol and other remote sites located at Birmingham. Short, Belfast, Northern Ireland, and AVRO, Chadderton, near Manchester, built Bristol Bombay aeroplanes and Rootes, Speke near Liverpool, built Bristol Blenheim aeroplanes that were first delivered to the RAF in 1937.

The Bristol Company continued to expand with the demand for greater production. The drawing office was enlarged, a new canteen added, and a recreational area, pavilion, and sports field completed the expansion. The Armament Section was created to design power-operated gun turrets and their associated hydraulic systems. The Armament Division would later become the Light Engineering Division responsible for cars and plastics. The stress section was coping with the new design of a mid-wing layout, a totally different set of problems.

Simultaneously in 1937, the Bristol Bolingbroke was being made in Canada by Fairchild Aircraft Ltd (Canada), Longueuil, Quebec, and the Bristol Beaufort made in Australia at Melbourne, Victoria and Mascot, New South Wales. The Beaufort and the Beaufighter would maintain Bristol's ability to build and develop their own aircraft. A derivative, the heavily armed Buckingham bomber, was too late for the European War and was completed as a transport or as a Buckmaster trainer.

During this company buildup to the Second World War, tragedy struck. Barnwell had designed a light single-seater aeroplane for his own use. In August 1938, he was killed on his second flight. The loss was a tragedy to his family, the Bristol company, and the country, as he was one of the best designers in Britain. Leslie Frise would assume the mantle of Chief Designer. Frise invented the Frise aileron for better roll control.

Barnwell's family tragedy continued into the Second World War. They had three sons that served with the Royal Air Force and all three were killed in action. The Barnwell family sacrificed everything for our freedom. 'We will remember them':

Pilot Officer John Sandes Barnwell of No. 29 Squadron RAF died aged 20 on 19 June 1940.

Flight Lieutenant Richard Antony Barnwell of No. 102 Squadron RAF died aged 24 on 29 October 1940.

Pilot Officer David Usher Barnwell DFC, RAFVR of No. 607 Squadron RAF died aged 19 on 14 October 1941.

What of the flying schools established in 1910? No. 2 Elementary Flying Training School (EFTS), temporarily No. 6 Flying Instructors' School, was located at Staverton, Worcester, and Yatesbury. No. 10 EFTS, established in 1936, was at Yatesbury, Weston-super-Mare, and Stoke Orchard. By 1942 when it closed, it had trained more than 2,000 ab initio pilots flying 104,000 hours on the De Havilland Tiger Moth. Some of the flying schools continued training RAF Reserves until 1953. The Bristol Wireless Flight trained over 18,000 wireless operators, flying 224,000 hours, at No. 2 Radio School, Yatesbury, flying the De Havilland Dominie (the civil de Havilland DH.89 Dragon Rapide) and Percival Proctor.

The Aircraft Service Department operated works in Scotland, Lancashire, Surrey, and Wales for 'factory repair' or, if irreparable, the wrecks were sent to Avonmouth, a suburb of Bristol, for salvage. Mobile repair parties also were sent out across the country as needed and repaired nearly twenty per cent of aeroplanes on site.

The first raid on Bristol by the Luftwaffe was on 24 June 1940. The Works were protected by a balloon barrage and anti-aircraft guns. During the Battle of Britain, the air raid sirens would sound causing some work interruption. On 25 September 1940, 100 bombs were dropped on the Works, killing 72 employees and injuring 166, of whom 17 later died. Many of the departments that did not have to be at Filton were dispersed to the surrounding area. In April 1942, a Luftwaffe bomb demolished the wind tunnel and adjacent office building. The Bristol conglomerate was now at its peak employment of 52,000 personnel and fulfilled all its obligations until the end of hostilities in 1945. These were the monoplanes that the Company designed until the end pf hostilities.

1934–1945

Bristol Prototypes/Demonstrators
Type 133 (1934) was the first cantilevered, retractable geared, fully cowled Mercury-engined, Alclad monoplane prototype offered to the RAF. Alclad is an American product of corrosion-resistant aluminium sheet created by bonding aluminium surface layers to high-strength aluminium alloy core material. The only fabric-covered components were the controls. Unfortunately, during final trials a spin was entered with the landing gear down and the aeroplane crashed. Production: 1

Type 142 (1935) was created out of remarks by the Prince of Wales – 'We must think in terms of cruising speeds of 250 mph' – and by Lord Rothermere, proprietor of the *Daily Mail*, who said that he wanted 'the fastest commercial plane in Europe'. This was to point out to the Air Ministry that their present fighters were too slow to match a high-speed light bomber. The Mercury-engined aeroplane, named *Britain First* by Lord Rothermere, created a furore in June when it proved to be 81 kph (50 mph) faster than the Gloster Gladiator fighter. It remained as an experimental aeroplane until 1942. Crew: 2, Passengers: 4, Production: 1

Type 143 (1936) was the slightly larger version of the Type 142 with Aquila engines. It never achieved success because a variable-pitch propeller was not available to fit the Aquila engine. Discussion around a 142M bomber variant with Mercury engines halted 143 development. Crew: 2, Passengers: 8, Production: 1

Type 138A (1936) In the race for manufacturer's prestige, Bristol decided to attempt the world's altitude record with a monoplane powered by a 373 kw (500 hp) Pegasus engine. It succeeded twice, once in 1936 and on 30 June 1937, with a homologated *Federation Aeronautique Internationale* record of 16,440 m (53,937 ft). In the ten years since 1928, the height record had been broken nine times, once by a Bristol Jupiter, and five times by a Bristol Pegasus. Fedden certainly could be proud of his engine achievements.

Bristol Bombers

Blenheim/Bolingbroke (1936) resulted from Barnwell's proposal to alter the successful Type 142 to a three-seater, mid-wing medium bomber, Type 142M, with a Browning gun and bomber's station in nose and a dorsal turret. By raising the wing it allowed for internal bomb storage. The Blenheim was exported to Finland, Turkey, and Yugoslavia and the Bolingbroke was built in Canada. It is beyond the scope of this book to tell the full story of the Type 142M and all it variations but suffice to say, its contribution to the pre-war buildup and initial years of the Second World War was tremendous. It was another outstanding success for the Barnwell/Fedden team at Bristol. Crew: 3, Production (all models): 6,215

Beaufort (1938) was a torpedo bomber and mine-layer developed from the Blenheim. It was made under licence in Australia as the DAP Beaufort. It used the Pratt & Whitney Twin Wasp and various models of Bristol Taurus engines. Crew: 4, Production: 2,080

Buckingham/Brigand/Buckmaster (1943) was designed as a fast medium day or night bomber. Too late for operational widespread use in Europe, and the war ended before deployment to the Far East. Crew: 4, Production: Buckingham 123, Crew: 3, Brigand 147, Buckmaster 112

Bristol Fighters

Type 146/148 (1936) were the last single engine aeroplanes constructed at Filton as the Company turned to multi-engine aeroplanes. Both aeroplanes were not successful in their bid for a contract, Production: 146 1, 148 1

Beaufighter (1939) see Chapter 6

Bristol Transport

Bombay (1935) was a high-wing monoplane designed to carry twenty-four fully equipped troops. The Bombay was an official contract on the Secret List so no aeroplane facts could be published. However, it had features which would be incorporated in the Bristol 170 Freighter ten years later. Bristol was fully occupied with the Blenheim and because of its physical size the contract was awarded to Short and Harland Ltd., Belfast, Northern Ireland. Production: 50

1945–1960

Post-war, Bristol set up a separate helicopter division under Raoul Hafner in the Bristol Oldmixon factory at Weston-super-Mare, Somerset. He was instrumental in developing the Type 171 Sycamore and the Type 192 Belvedere for the RAF. Hafner was more interested in civil applications of rotor craft but this was stymied by the 1960 government-enforced merger of all helicopter companies under Westland Helicopters.

Bristol Cars was another post-war project based at the facility in Patchway, Bristol. The engine that powered the Bristol 400 car powered other manufacturers' cars such as Cooper, Frazer Nash, and AC. An overseas version, the Arnolt-Bristol, was a sportscar sold in the US.

Peacetime expansion at Filton included a new design office on the site of the bombed wind tunnel, an eight-acre Assembly Hall, and a lengthened, widened, and strengthened main runway to accommodate larger aeroplanes. In the fifteen years following peace and the formation of the British Aircraft Corporation, Bristol's interests included the Type 170 Freighter/airliner, Brabazon and Britannia commercial airliners, investigating supersonic flight with the Type 188, and the Bloodhound Missile. In 1959, Bristol was forced to merge its aircraft interests with English Electric, Hunting Aircraft, and Vickers-Armstrongs to form the British Aircraft Corporation conglomerate.

Bristol 170 Freighter see Chapter 6

The Bristol Brabazon was the result of the Brabazon Committee's specifications for a large fuselage, luxury (large personal passenger space), transatlantic airliner with amenities such as a cocktail bar, cinema, dining room and sleeping berths. Eight Bristol Centaurus engines provided the power to eight paired contra-rotating propellers. Unfortunately, the package was not attractive enough for the airlines because of its high cost per seat mile. What it did do was accelerate aeroplane system development by Bristol, which would be used on future aeroplanes. This would include fully powered flying controls, electric engine controls, and high-pressure hydraulic systems. It first flew off the lengthened runway at Filton on 4 September 1949 and subsequently at the 1950 Farnborough Airshow and 1951 Paris Airshow. The largest aeroplane of its time, it was scrapped in 1953, only the one prototype ever flew.

The Bristol Britannia was designed to fly long distances throughout the Commonwealth. It just foreshadowed the arrival of the pure jet aeroplanes, consequently only eighty-seven were built, but it was successful in the interim with its turboprop engines. Initially designed with the Centaurus engine, the prototype went airborne with Proteus engines on 16 August 1952 followed by an appearance at the Farnborough Airshow the same year. Engine teething problems caused further delays until an in-service date of 1957 which was too late for the domestic market, the de Havilland Comet had gone into service with BOAC in March 1952, albeit with serious issues.

However, Canada was interested in the Britannia, not only with Canadian Pacific Air Lines, but three versions were produced by Canadair. The Argus was a piston maritime patrol aeroplane that used certain Britannia components, the turboprop CL44 was a freighter aeroplane with an innovative hinged tail section for ease of cargo loading, and the turboprop Yukon was a military transport. They both used Rolls-Royce Tyne turboprop engines.

In the mid-1950s, Bristol constructed the experimental Type 188 to examine flight at Mach 2+. The aeroplane would have to use new experimental metals due to the heat generated by the friction of the supersonic speeds. The Type 188 first flew on 14 April 1962, now as part of the British Aircraft Corporation. Also during the mid-1950s, Bristol was developing the Bloodhound missile. The Mach 2+ missile had two ramjet engines and four solid fuel booster rockets and was classified as a defensive surface-to-air missile. It entered service with the RAF in 1958.

The engine division was renamed Bristol Aero Engines in 1956, which then merged with Armstrong Siddeley to become Bristol Siddeley in 1958; ironic, as the Bristol Jupiter was the direct competitor of the Armstrong Siddeley Jaguar in the 1930s and proved itself superior.

In fifty years the Bristol Company had come from a pusher type biplane to the world's largest aeroplane in 1950, to developing the Mach 2+ Type 188, from the three-cylinder Lucifer to the mighty and most powerful eighteen-cylinder sleeve-valve Centaurus engine, and facilities that started in a tram shed to, by the mid-1930s, the world's largest aviation production facility.

So perhaps some numbers, rather than words, will aid in further understanding the production accomplishments of Bristol: 15,750 aeroplanes of eighty-five designs and 8,320 aeroplanes of Bristol design built by licensees and contractors.

The extended White family built a company that Britain could be proud of and the Bristol company's contribution of engines and aircraft in wartime must be recognised for what it was, a tremendous effort which contributed to the Allied victory. Further post-war contributions were made to the British aviation industry until 1960 and then subsequently as part of the British Aircraft Corporation that produced the supersonic Anglo-French Concorde. The company legacy is preserved at Aerospace Bristol on the historic Filton airport and The Helicopter Museum at Weston-super-Mare for present and future generations.

2

SIR ROY FEDDEN

In this chapter, I wish to acknowledge extensive use of The Rolls-Royce Heritage Trust's Historical Series No 26, *Fedden – the Life of Sir Roy Fedden* by Bill Gunston, OBE FRAeS, which provided information and assisted me to confirm other sources.

Fedden went against all society and family expectations to become an engineer. Thank goodness for the world that he did make that choice, first, because of his contribution to the Allied war effort in both the First and Second World Wars, and second, because of his contribution to civil and military aviation post-war. His contribution to the design and production of aero-engines must never be forgotten, and it is the author's hope that this book will help preserve that memory.

Alfred Hubert Roy Fedden, later Sir Roy, was a larger-than-life character who demanded the proverbial 110% from himself all the time. He expected the same, without reservation, from all his employees during work hours and often outside them, as required. He could be described as a benevolent engineering dictator, light on the benevolent and heavy on the dictator, driven to lead and control all innovation, performance, and results. It was all for the good, which is what prevented him from crossing the fine line and being impossible to work with.

We shall see, however, that his very successful engineering empire would eventually be his downfall. Some people did not appreciate his style of management. Pity, he had a lot more to give. Unfortunately for Fedden, some of those people sat on the Board of the company that he worked for, the Bristol Aeroplane Company. Here is his remarkable story, his achievements, his contribution to the British aviation industry, and his legacy. His many aero-engines were heard around the world.

Lord Kings Norton included these words in an address at the memorial service for Sir Roy Fedden at the church of St Clement Danes, London, on 6 February 1974: 'His views were convictions, firmly held and invariant. He was tireless in pursuing the ends he believed in. His integrity was absolute.' Fedden's life was not plain sailing. He rode the peaks of success and recovered from the troughs of failures and disappointments.

The Feddens had been in the Bristol area for nearly 300 years and although a wealthy family, Fedden's parents, Henry and Mary Elizabeth, were not immune to the poverty surrounding them. His mother was the daughter of a Methodist clergyman. They both had a fine sense of social responsibility and addressed the social problems evident in parts of Bristol. What could they do to improve the living conditions of some of the citizens?

Henry believed training in a trade could help. He utilised HMS *Formidable,* one of the last ships of the line, to train penniless and orphaned boys at Portishead. It was a tough training course, but it gave the boys a chance to get a start in life. Many of them would go on to serve in the Royal Navy. He started a local chapter of the Society for the Prevention of Cruelty to Children, was President of the Bristol Lifeboat Association, and a trustee of The Hall School at Stoke Bishop. Into this caring family Roy Fedden was born on 6 June 1885, the third son, and grew up in *Fernhill,* a large house in the village of Henbury close to Bristol. The house had a large garden and Roy could be found in his youth happily assisting the gardener and his boy assistants from the village. He gained an appreciation from them how the other half lived.

The Hercules: The Other Engine That Helped Win the War

He went to a public school and a private school, but continued to enjoy gardening, fly fishing, and carpentry in the workshops at *Fernhill*. This was the beginning of his interest in fashioning and creating things, in design, and in construction. On a family holiday to Porlock he unfortunately had his hand caught in a closed door that took many years to heal. On another trip through Somerset with his father he gained a further appreciation of how the ordinary working people lived and their views and aspirations in life. Bill Gunston in his book, *Fedden – The Life and Times of Roy Fedden*, quotes Fedden: 'My sympathies with the underdog and young people struggling to get established often caused eyebrows to be raised. But this outlook earned me respect, affection and loyalty that money could never buy.'

The Fedden family were living in the last of the horse-drawn era, but that would change. Roy went off to Clifton College, which upheld the traditions of the elite public schools. There was arduous, rigid discipline, and the very British 'play up and the game' comes from a poem by Sir Henry Newbolt that refers to a cricket match at the school. He was in the military side of the school but was rejected by the Navy and Royal Marines, so he had serious doubts about a Service career. Then something happened that would have a profound affect on his life's direction, his Dad bought a motor car.

Vincent, his brother, was tasked with bringing the car back to Bristol from H. M. Hobson Ltd in London. Roy would join him in Reading for the second half of the journey. The car was an 8½ horse-power, two-cylinder Decauville. It was, they discovered, not very reliable and underpowered. Roy rolled up his sleeves and often took part in the repair of the car. It was eventually traded for a Bristol Motor Company car. About this time, his Headmaster at Clifton College recommended him as a Sandhurst candidate for a career in the Army. Just prior to departure, he made a momentous decision. He would not be going to the Army. The Decauville had won. He announced that he wanted to become an engineer.

This came as a shock to his father, that one of his sons was not going to carry on the Fedden family tradition of socially acceptable careers. Engineering was not one of them. Family and friends begged him to reconsider, letting the side down old boy, but Roy instead remembered such people as Stephenson and Brunel who had led the Industrial Revolution. Roy was lucky, he had an understanding father in Henry who, when he realised the decision had been made, would support him in his career ambitions. Roy Fedden would become an engineer.

Once again, Fedden was lucky, as he enrolled in a local paid apprenticeship program with the Bristol Motor Company during the day, and at night studied at the Merchant Venturers' Technical College. He was not looking for a superficial education like others of his social standing, but really wanted to know how to create and fix things. Donning overalls, he took part in testing engines and materials, operating lathes and drill presses, and the art of hand forging, heat-treating, quenching, and grinding. Nothing was left out in his bid to understand the processes involved. That was the practical side. He also devoted his spare time and energy to motor car design in 1906. To his mind, the motor car was here to stay despite society's reluctance to accept the noisy, smelly vehicles that frightened the horses. He was introduced to John Brazil, a qualified engineer and leading businessman. The social connections did work...

In 1907, Fedden approached Brazil, now of Brazil Straker & Co., with his design of a two-seater car aptly named the *Shamrock*; Brazil was Irish. Not only did Brazil consider the design but he engaged him as a junior draughtsman, the wheels of his illustrious career were rolling. In November, the twenty-two-year-old exhibited the car at the motor show at Olympia. It had many unique Fedden features, which he and a team of engineers had designed and produced. He himself was there to explain its features. The cheap four-cylinder car was an overwhelming success. Roy Fedden was made chief engineer of Brazil Straker.

The leader of Fedden's team was chief designer L.F.G. 'Bunny' Butler and this partnership of Fedden/Butler is at the very heart of this narrative. Theirs was a bonding of ideas, design, and engineering competence resulting in an exemplary product.

The Hercules: The Other Engine That Helped Win the War

Butler went to Crossley Motors in Manchester for a short time, returning as chief draughtsman in 1912. The scene was set for the dynamic duo to impress not only the motor car industry, but subsequently the aviation industry. In 1908, the Straker Squire car took the motor show by storm, setting records in its category. 1908 to 1911 saw the car further developed with achievements on the Brooklands circuit and test hill. Fedden was a keen driver at these events.

The year 1911 was not so good. He had to lay off nearly 300 men due to a lost contract, and he injured himself playing football. He left Brazil Straker for a year to recuperate in the warmer climate of South Africa. He returned as chief engineer at Fishponds. He saw the future of motoring, but flying was more and more in the public eye. In 1912, he attended the Military Aircraft Trials on Salisbury Plain and the flying activity at Hendon and Brooklands. Brazil and Fedden went to France seeking aviation business opportunities, the Clerget engine being of interest.

In 1914, Fedden was appointed technical director with a seat on the Board. There were no suitable British aero-engines available as war approached, so the War Office organised a competition. Fedden examined the submitted engines and was impressed by the Argyll submission using a patented Burt and McCollum sleeve-valve engine. It won the competition but got lost in the flurry at the start of the First World War. During the same period, he went to Stuttgart to get the Bosch electrical system for his cars. While at the Mercedes factory, he inadvertently discovered an in-line, six-cylinder, water-cooled engine that was destined for aircraft. This concerned him, but it did open a door to further thought.

Fedden tried to enlist but was turned down because of his old injury. The Fishponds works were growing quickly, with 2,000 workers, increasing numbers of female workers, and delivering staff cars, lorries and artillery shells to the War Department. Then came another major turning point in Fedden's life. Commander Wilfred Briggs, his sailing friend, Head of the Royal Naval Air

Service (RNAS) Engine Division, wanted to assess Fedden and Brazil Straker to see if they had the capability to cure problems with the widely used 90 hp V-8 Curtiss OX-5 in the Curtiss JN-4 training aircraft. The engine was causing too many training accidents. He was sent 300 crated engines for examination. He quickly concluded that a re-design was required. In two months, the engine was re-designed, manufactured, and now had a 200-hour life. The RNAS training efficiency was transformed thanks to Fedden and his team.

Brazil Straker became the only manufacturer allowed to build Rolls-Royce engines under licence in the First World War. The first engine made was the Rolls-Royce Hawk. The second complete engine was the Rolls-Royce Falcon for the Bristol Fighter, followed by components for the Rolls-Royce Eagle, the bigger version, and the French Renault WS engine. The company continued with their original manufacturing agreement for vehicles and ammunition. By 1916, the water-cooled engines were the only ones producing the required horsepower, the air-cooled engines did not keep up; but the water-cooled ones were susceptible to water leaks owing to both war damage and inadequate manufacturing tolerances. Fedden pondered this dilemma at length.

In 1917, the Air Department at the Admiralty drew up a specification for a static air-cooled radial with diameter less than 42 in (107 cm), weight less than 600 lb (272 kg), and maximum power greater than 300 hp (224 kw). Fedden and Butler threw themselves into the challenge, Fedden the designer and Butler creating the engineering drawings. FB was put on all drawings and was the prefix for part numbers. They created the fourteen-cylinder, helical configuration radial Mercury engine. The ABC Dragonfly won the competition and turned out to be a classic failure among aero-engines, but that is another story. That year, his father died and Fedden became involved with an older, divorced lady with a son. They later married and set up home at Widegates, Westbury-on-Trim.

Fedden realised that the Mercury was good for the present time but would have to improve greatly for the future. His ruminations over, Fedden decided on air-cooled radial engines and designed the nine-cylinder, single-row, air-cooled Jupiter engine. It ran in October 1918 just a month before the war's end. However, company trouble was brewing. Brazil left for new business ventures and the works was bought by an Anglo-American financial group – interested in shipping and coal!

Cosmos Engineering Co

Fedden was instructed, by letter, to carry on as before. The future did not look bright for the company, in spite of its wartime success. He did, however, receive a letter from the Ministry of Munitions of War to continue with his engine development. Simultaneously, the Fedden/Butler duo were already designing an unusual three-seat, three-cylinder, air-cooled engine motor car. Once again, his design was a success at the motor show in Olympia, with 2,000 orders. In 1919, the car part of Cosmos was sold to Straker Squire Ltd. A personal team success, but his heart, I suspect, was still connected to his achievement with the Curtiss OX-5. A little 100 hp, three-cylinder engine called the Lucifer was now running on the test bed.

The company development started to lag as there were no funds to develop the Jupiter. He knew he was being head-hunted by various aviation firms, but he enjoyed the total control he had at Cosmos and remained at Bristol. The outlook for Cosmos was bleak as the post-war period began. The wartime contracts dried up as the country suffered under its huge debt. However, one thing showed promise, Fedden's engines performed well and he was capable of developing them further. Towards the end of hostilities, Captain Frank Barnwell, chief designer at the British and Colonial, later Bristol, Aeroplane Co. was looking for a replacement engine for their Bristol Scout Type F. With the Cosmos Mercury engine installed, this would become the Scout F1 or Bristol Type 21A. The Royal Aircraft Establishment, previously Factory, at Farnborough, tested the combination in April 1919 and the Mercury-Scout set climb-to-altitude records.

Almost simultaneously, the second Cosmos Jupiter engine had been installed in a Bristol Badger. It performed well, despite a rudimentary installation for test purposes. The third Jupiter was installed in a Sopwith Schneider seaplane with impressive speed results. Later in 1919, the Lucifer was installed in an Avro 504K test-bed aircraft and an aircraft was ordered from Westland to be the test-bed aircraft for the Jupiter engine. Then Fedden's engineering world came to an abrupt halt. Cosmos went into liquidation due an overseas venture gone wrong. Fedden was now holding the Fishponds Works for the Receiver. This must have felt initially catastrophic for such a driven, ambitious individual.

Bristol Aeroplane Company

Several companies were interested, but for various reasons did not pursue the matter further. Fedden and the Cosmos team had proven themselves to be a valuable asset to Britain and their engine exports promised to help the nation's recovery. The Air Ministry put pressure on the Bristol Aeroplane Company (Bristol), who cannily professed no interest in Cosmos, to purchase the entire company, which they eventually did at a bargain price – probably a quarter its value. The Bristol Board established, on a trial basis, its Engine Department on 29 July 1920. Fedden immediately proposed development of the Jupiter to 500 hp and asked for £200,000. This was approved by the Board with the understanding that if the proposal failed, the Engine Department would be closed. It was all up to Fedden and his team to prove their investment was worthwhile.

The mainspring in all that would follow was Fedden – not how Fedden would handle the White dynasty, but how the Whites would cope with the engineer Fedden, their employee. He was a larger-than-life figure, a forceful personality, who would certainly not kowtow to the White family where his engines were

The Hercules: The Other Engine That Helped Win the War

concerned. In fact, he had just been awarded an MBE, Member of the Order of the British Empire, for his wartime work.

The Board were in agreement that Fedden was important for potential for new business, but that he could never be appointed to the Board. He was that engine man down at Fishponds. Socially, the Feddens had been around Bristol for a hundred years longer than the Whites and certainly had no need to feel excluded from the 'old boy's club'. Could they work together to make the investment work? Could they communicate, the visionary experimental engineer and the non-technical businessmen looking after the interests of the stockholders and themselves? Could they make the engine/airframe combination work? It would be difficult, but possible.

The whole aviation industry was struggling due to the inventory of wartime surplus aircraft and engines. Fedden and his team of thirty-two workers were surely concerned about their future when an opportunity, some would call it a breakthrough, occurred. On very short notice he was instructed to prepare Lucifer, Mercury, and Jupiter engines for exhibit at the first Society of British Aircraft Constructors (SBAC) exhibition on 20 July 1920. Incidentally, the SBAC was founded by Sir Henry White-Smith in 1917. There were mixed reactions to the 'new kid on the block' putting up such a display, but Fedden welcomed the opportunity of the public display. Work was scarce and what work there was should include one of his engines. The inhouse Bristol Bullet racer had a Jupiter engine installed and came second in the Aerial derby. In August, the Engine Department moved from Fishponds to the corner of Filton Aerodrome, former Air Ministry property, near the village of Patchway. He was conveniently right among the aircraft and engines, ideal to his mind.

The new premises were emptied of wartime debris and provided with gas, electricity, and water. Fedden's thirty-two strong team would require a few partitions in the large shops. They worked long hours to prove the concept could work for Bristol. They were proud of their engines, they wanted to beat their competitors, they wanted to secure their own personal future amid the dearth of aviation contracts. They also had a leader in Fedden who worked even harder than they did and believed that the Jupiter was cost-effective. The Aero Engine Department had the Air Ministry order of ten Jupiter engines to fulfil, but nothing more after that.

Fedden had the competitive advantage in that he was creating a lean business organisation from scratch within BAC. The post-war economy allowed him to pick up machine tools and outsource some complicated work cheaply. The design, development, and manufacturing process was very controlled. The envy of cumbersome established companies, he was approached by them with offers of employment. However, he had his engines, his team, his hand-crafted facilities, and he was in Bristol with all his connections. The next challenge was to sell his product, the hard part in post-war Britain.

He spent his spare time writing about, and preparing lectures on, the air-cooled versus water-cooled engine debate. Fedden was firmly in the air-cooled corner. The legacy of development of aero-engines after the First World War left unanswered questions about the best type of engine for the application, inline or radial, air- or liquid-cooled? The two types of engines each had their advantages and disadvantages, and it really became a balancing act for the aero-engine manufacturers between their previous engine experience and the government engine requirement. Power, weight, efficiency, and reliability were the demands on the internal combustion engine, the propeller was another matter.

The central theme of engine building was maximum horsepower for minimum weight, overbuilding components was not allowed. In previous wartime years, engine parts would frequently fail, Fedden wanted to know why. The word 'crystallisation' was used as a general reason. He engaged a metallurgist Professor Aitchison to test the raw material for his Jupiter engines. Most of it was below standard. Aitchison then went on, with Fedden and Butler, to examine the mechanical design of the Jupiter and discuss the new concept of fatigue that could occur from the smallest, microscopic stress concentration.

The Hercules: The Other Engine That Helped Win the War

The answer was smooth polished surfaces, no sharp corners, blended joints, generous curves: in other words, attention to detail. Fedden then had a fatigue testing machine set up to test all his parts to get some idea of their expected and safe useful life. To improve quality control, he went to the suppliers to explain what was now required of them. Far from expressing resentment, they realised that this was a new age and they had better adapt to remain in business. Fedden set up the department's own foundry close by on the aerodrome to ensure quality. His overall aim was to control his supply line from start to finish, no weak links allowed. One perceived as such was lack of qualified engineers. Fedden addressed this issue by establishing an apprentice program within the Engine Department. The engineers produced were sought after by other companies upon graduation, such was the program's reputation.

Fedden had great plans for the Jupiter. He wanted to change the split master rod to a single rod running on a two-piece crankshaft and use improved expansion compensating valve gear. The Jupiter II passed the Air Ministry type test in 1921, the first air-cooled engine to do so. Then the dreaded call from the Board occurred; the development money for the Jupiter was all gone and unless something changed, namely sales, the Engine Department would be closed. Fedden, in a last-ditch sales effort, took the Jupiter engine, 1.4 lb/hp, to the Paris Airshow. Saved at the last hour by the French Gnome-Rhone company acquiring a manufacturing licence, BAC received a large sum of money.

The French factory left a lot to be desired as compared to Bristol, but it still insisted that the engine passed the Service Technique standard, which it did in June 1922. In fact, in 1923, a 436 hp Jupiter IV passed the 100-hour test with lower levels of wear never achieved before. This same year was the turning point for the Engine Department and BAC when the Air Ministry ordered eighty-one Jupiter IV engines. At age forty, by way of Brazil Straker, Cosmos Engineering, and the Bristol Aeroplane Company, Fedden had finally achieved at least the beginnings of his ambitions, with a promising engine and the works in place to develop greater engines for the future.

All of a sudden, his engines were installed in different aeroplanes. Fedden formed an installation department under Freddie Mayer to cope with the demand. The first Service aircraft to install the Jupiter was a Nieuport Nighthawk. This was followed by installations in specially built Bristol Racers, Hawker Woodcocks, and the Bristol Fighter, which went to Sweden. Barnwell adapted the Bristol Fighter to a two-seat trainer used by the Bristol Flying Schools. Fedden's Lucifer engine was now winning Aerial Derbies and, in the public eye, it was now making revenue from overseas. The small Cherub flat-twin was very successful in light aircraft competitions. A monoplane with a Cherub engine flew from Munich to Rome reaching altitudes above 14,000 ft crossing the Alps. The pilot's name was Willi Messerschmitt.

The Jupiter was installed in a French racer and established a world record on a closed course. In response, the French designers included the Jupiter engine in all their new aircraft designs, civil and military, from 1923 until 1929. Fedden had established two important work areas, one was for examining all failed parts and the other for examining all manufactured parts. In 1924, he hired F 'Pop' Nourse from Daimler to be a Quality Manager, a new concept at the time. Frank Owner ran the rig testing section, which included Fedden's idea of a one-cylinder testing rig. Design, performance, and reliability were the buzz words. This all led to engines becoming more difficult to make. Fedden headhunted Fred Whitehead, who he had worked with him at Brazil Straker, to solve this problem, and he did.

For Fedden's team's ultimate test, six Jupiter engines were run in front of the AID Inspector and then dismantled. The parts were all mixed up, the engines reassembled, and they all ran perfectly. The final proof of performance for the Jupiter was two flights, one was a 225-hour shuttle flight, and the other was an England to Cairo return flight. The Jupiter had confirmed that it would be the most important engine for the next ten years with both

The Hercules: The Other Engine That Helped Win the War

the Royal Air Force and Britain's civil flagship carrier, Imperial Airways.

Building on this success, Fedden tackled the problem of losing power with altitude caused by the thinner air. He examined higher compression ratios, variable inlet-valve timing, and supercharging, both mechanical and turbo. Some of his results on turbocharging were shelved due to lack of money. After a 1925 visit from some Wright Aeronautical personnel, Fedden went to the US on the first of many visits. He became well known and was well respected among his American friends and rivals in Pratt & Whitney and Wright Aeronautical. He was particularly interested in their starting systems and variable-pitch propellers.

In 1927, Fedden developed what he regarded as a new family, the Jupiter VIII with a geared propeller. He also designed the Titan to fill the power gap between the Lucifer and Jupiter engines. This was followed shortly by the 315 hp Neptune engine. The Jupiter was now licensed in seventeen countries including Bristol Engines of Canada, Montreal. These agreements were worth a million pounds royalty annually. It appeared that all Jupiter engines would be made outside Britain until the RAF ordered engines for 312 British Bulldog fighters and 505 Westland Wapiti multi-role aircraft. The Engine Department would be busy and financially sound.

To streamline the cylinders of the radial engine a 'Townend Ring' was fitted around them. A young engineer suggested to Fedden that the exhaust be used as added thrust. His name was George Dowty. There was no stopping the Mark XI Jupiter-engined aircraft, from altitude records to long distance biplane and seaplane flights, from multi-engined installations to four-bladed propellers. The Jupiter was at its zenith powering fifty per cent of the world's airlines and air forces. The team had grown to 1,900 under Fedden's total control. He had made Bristol the proverbial 'mint'. Why was he not offered a directorship? Perhaps the White family was afraid of losing control, or future conflicts, and remember this was 100 years ago. there were probably social misgivings, as after all, he was only an engineer!

Sir Roy Fedden

By 1930 he had been with Bristol for ten years, but only had one major product. He was researching four areas for another one, the double-row radial engine, the diesel engine, fuel injection, and the sleeve valve engine concept. He did have a programme for split hub metal propellers in place, but there was no money for variable-pitch research. What would he want to do that for? His ambition was later realised when Rotol (propellers) was established in 1937 with a Bristol and Rolls-Royce partnership. These propellers later gave Hurricanes and Spitfires extra speed, all thanks to Fedden's perseverance.

Personally, things were not going well at home, too much time at the office. So much so that Mrs Fedden found someone who would give her that companionship and, when caught by Roy, disappeared from the home scene. Widegates was then run efficiently by a housekeeper. The staff would be invited to dinner during the week and after the meal, the charts would come out to be discussed. During the 1930s he re-established contact with a former girlfriend from his youth, Norah, whose husband would not entertain a divorce. They maintained a discreet relationship for many years.

By the early 1930s, the Jupiter was at the end of its development and the Pegasus and Mercury were the new focus. In 1931, Fedden was working with direct fuel injection into the cylinder and was convinced of its future. The Air Ministry was not, and the concept was dropped. A huge mistake on their part. Germany progressed with the idea. Fedden was also working on a double-octagon engine called the Hydra, which achieved 870 hp (649 kw). Something then happened in 1933 that would greatly affect Fedden and Bristol (and some others), something that was out of their control. Adolf Hitler came to power in Germany.

In 1934 Fedden took the apprenticeship scheme to the next level by establishing the Bristol Engine Apprentice School. He also adapted the Mercury and Pegasus engines to the new DTD.230/87 Octane fuel with tetra-ethyl lead. The engine royalties were increasing his wealth at a great rate, another cause of concern for the Board. These extra funds allowed his three passions to flourish: fishing, motor racing, and power boat racing. All were approached, not surprisingly, with the attitude of winning!

In 1934 he gave a presentation on yachting where a representative from the newspaper tycoon Lord Rothermere approached him and asked if he could build a faster modern transport than the Americans. Fedden said yes. The type 142, *Britain First*, flew on 12 April 1935 with Mercury engines and achieved speeds exceeding the fastest fighters by 80 mph (129 kph). Bristol got an order for 150 variable-propeller, Bristol T142M Blenheim monoplanes. This was all in addition to developing Pegasus engines for the Short Empire flying boat.

Fedden had researched and pondered the sleeve-valve engine for some time. We shall see details in the following two chapters of the results of his development of the Perseus, Aquila, Taurus, and Hercules sleeve-valve engines. By 1935, he was trying to awaken the government to the German threat but to no avail. He was aware of their industrial might. The powers-that-be thought he was grandstanding. In 1936, the ill prepared industry was working flat out before it reached peak production years later. Bristol was turning out Mercury and Pegasus engines to be joined by the Perseus engine. Fedden had discovered that Germany was building engines as complete 'power eggs' for easy replacement.

Eventually, Sir Hugh Dowding supported him and shadow factories would be established to continue production of Bristol engines away from the main facility in the event of a major aerial attack. These factories were run by car companies such as Austin, Rootes, and Rover and would employ over 51,000 workers. In 1937/8 the Bristol 138A monoplane established altitude records and distance records, Egypt to Australia. Both used Bristol Pegasus engines. Fedden was also working on a high-altitude version of the Hercules engine for the Vickers Wellington V, but this was dropped when the war started.

Shortly before the war, in 1938 various German delegations visited the British aviation industry and Fedden went on reciprocal visits. On one visit he was honoured and awarded the

Lilienthal Ring by Adolf Hitler. He was amazed at what he saw, the technical advancements and development work at BMW on a gas turbine. He wrote comprehensive reports on his observations, to no immediate effect. He outlined a whole series of measures that should be undertaken straightaway, from the types of aircraft required to the structure of the government, to make Britain's response work effectively. Was all this German transparency he saw to dissuade Britain from opposing the Nazi expansion? It had the opposite effect on Fedden. It made him more dedicated to opening the eyes of the British Government, despite being branded a scaremongerer.

In August 1939, the Engine Department offices were spread out to various locations. The Design and Project Office moved to Tockington Manor, and the Main Design Office to Somerdale. The Engine Department had grown by now from 31 workers to 16,600. Fedden suggested to the Board that the company be divided into two divisions to make it more efficient. Really, what did an engineer know about running a company? He never got a reply. The Hercules engine was having some problems and demanded attention and Fedden forgot about the letter. The Centaurus engine was sidelined temporarily because the emphasis was on production, not development.

Norman Rowbotham had returned from France to become engine works manager. A different personality, he was gaining favour with the Board, the Shadow factory owners, and the government officials. At the end of 1939, Fedden was called in by the Chairman of the Board and told that his engine royalty payments could not continue at the present rate; he was obviously too successful! He accepted the reduced commission, but as the Board refused to discuss his reorganisation letter, he said that he would work for the duration of the war and six months afterwards.

In Spring 1940, Fedden went to Canada to investigate the possibility of building the Centaurus overseas with some help from American technology and personnel. It never came to fruition. The Minister of Aircraft Production, Lord Beaverbrook,

The Hercules: The Other Engine That Helped Win the War

recalled him to Britain and offered him a job as his assistant. He turned it down and remained at Bristol with 'his' engines. He would do the same thing again a year later, when prime Minister Churchill offered him a post overseeing all of Britain's engine production.

In September the war got more personal, not only was the factory attacked with loss of life, but Widegates, his beloved home, was demolished. He gathered what was left of his possessions and he moved to the office at Tockington Manor. In early 1941 Butler was working on drawings for the most powerful engine yet, the 4,000 hp (2,983 kw) Orion. This was the engine that Fedden thought would take them into post-war aviation. He was also working on a small turboprop, yes, after saying that he was not interested. It is hard to keep an engineer away from a challenge!

Rowbotham had been bombed out and was invited to live with Sir Stanley White in the interim. Fedden was slowly being politically outmanoeuvred within Bristol with the tacit approval of the Board. Rowbotham was obviously the chosen heir apparent. How, and when could Fedden be replaced? On 1 January 1942 he received the news that he was on the Knights Bachelor Honours List. He was now Sir Roy Fedden. Butler threw a cocktail party to celebrate. The Board said nothing for a few days. How could an engineer be one of them? The Board were in shock as they considered this turn of events. It was too much. In February, Fedden received a letter from the Chairman of the Board giving him six months notice because they were dissatisfied with his handling of the executive side of his department.

Fedden continued to work. He was particularly interested in a BMW 801A engine from a downed Dornier bomber that had been brought to the factory, and the engine installation of a Fw190A-3. He copied one idea, the rear-facing exhausts to the Centaurus, which was now coming in to its own. Fedden finally got his underground bombproof factory at nearby Corsham with 4,000 workers. Many of them were women in the production department.

The Board was interfering more and more with proposed new and unworkable agreements, all to the detriment of Fedden's work, which they wished him to sign. The Minister of Aircraft production wanted him to continue his work until the end of the war, but the Board resisted government 'interference'. On 1 October 1942, the Board sent Fedden a letter to stop work, leave his desk as it was, and vacate the premises. True to form, Fedden held a sixteen-page agenda design and research meeting that evening for the future of 'his' engines; dedication or what?

Post-Bristol Aeroplane Company

I can just imagine the shock, dismay, anger, and disbelief of Roy Fedden and the entire Engine Department that he had built up over twenty years. To take this dramatic step in the middle of the war, after all that Fedden had done and was still doing for the wartime engines, was a decision that not only could have ramifications for Bristol, but possibly the country. A 'brave' decision, one that the Board felt necessary and justified. On the other hand, Sir Stanley White heaved a sigh of relief (no more confrontation, problem solved) and immediately praised the work of the new chief engineer, Rowbotham, and promoted him to the Board. The Board then established two separate companies, Bristol Aircraft Ltd and Bristol Aero-Engines Ltd. This was exactly what Fedden had proposed in his letter to the Board.

Fedden still had his engineering accomplishments, which could not be taken away from him. I think this would be most important to him, and the other stuff was just one of life's many disappointments. What was the cause of this dissension? Perhaps it was communication. The Board understood the greater picture and Fedden was so totally focused on the engineering challenges that he did not think of it as a business that must follow certain practices to exist. It was perhaps that he was dramatically too intolerant of any interference with his work that the Board eventually found him impossible to work with, despite all his successes.

The Hercules: The Other Engine That Helped Win the War

Fedden was nearly sixty, but with his reputation, the choice of jobs would be his. He chose to be Special Technical Advisor to the Minister; he was now part of the government. He was also chairman of the Royal Aeronautical Society Advisory Committee to the Minister. In December 1942, he led a team to the US to gather the latest technical information. Upon his return he submitted his *Mission to America* report, which, although not at all received well, at least got the industry talking. He noted three aircraft in particular: the North American P-51 Mustang, Douglas A-26 Invader, and the Boeing B-29 Superfortress. His expertise straddled the Services. He offered suggestions to improve Royal Navy torpedoes and Motor Torpedo Boats.

Fedden continued the success of his Bristol Apprenticeship program when asked to establish a College of Aeronautics. Educated aeronautical engineers would be in demand for future aviation companies. It would be eventually established at Cranfield. In 1943, the idea of Roy Fedden Ltd was formed to capture the expected post-war car boom. At the end of 1944, he led an investigative team to Italy to see what items 'would be a help to the allied war effort' and, upon return, he then was tasked with investigating engines required for emerging civil airliners.

In May 1945, he led his last investigative team to the rubble of a defeated Germany. His job was to gain knowledge on fuel injection, variable pitch propellers, and pick up anything that would be useful to the new College of Aeronautics at Cranwell. It was certainly slim pickings, as the Americans had scoured the earth for anything not nailed down. He still managed to bring back a selection of engines and a Volkswagen car: 'It will never sell,' they said.

Out of the government, he now turned his attention to his company. The sleeve-valve engine would propel the car, but was it the right choice? A few indecisive moves resulted, four- or six-passenger. Then a few crucial engineers left his employ and he had problems with the torque converter/transmission combination. Simultaneously, he embarked upon a flat six engine configuration. The company employed a large group of engineers, and its overheads grew very quickly, but nothing had been sold so far. Then, Fedden embarked upon a jet engine called the Cotswold. Was he trying to prove something to Bristol, or British industry, or maybe himself?

The swing-axle concept on the car proved unworkable, proved by track accidents. The engine work seemed to go in all directions with variations in size and configuration, achieving little financially. His turbo-prop project was abandoned by the Ministry of Supply. All were promising, but with no funds to bring them to fruition, Roy Fedden Ltd was liquidated in June 1947. He felt personally responsible for his staff and spent all his time finding them suitable employment, unusual and commendable. Then Fedden retired, age sixty-two. What would retirement look like for the master of British piston aero-engines?

He had builders extend the Buckland Old Mill, his long-time fishing hole on the river Usk in Breconshire, Wales. He did a little ocean racing with his yacht *Pegasa*, which he kept at Lymington, Hampshire. In 1948, he quietly married his childhood sweetheart Norah Crew, now Lady Fedden. They had twenty-five happy years together. One of his sailing compatriots was Henry Spurrier, later Sir Henry, who invited Fedden to be on the Board of Leyland Motors. It was not long before Fedden drew to their attention the fact that their engine was only rated at 140 hp (104 kw). He was of the opinion that a 240 hp (179 kw) engine would be required to remain competitive. His empire was starting to grow again in a different company, but he was outmanoeuvred politically; he quit this time, in 1949.

In October he was offered and accepted the newly created role of Aircraft Advisor to the North Atlantic Treaty Organisation. He and Nora moved to London and lived in their flat in Wellington Court. The work was all-encompassing and exhausting, all done with the highest level of security. The headquarters were then established in Paris, and the Feddens moved to an apartment overlooking Longchamps racecourse. The weekends often included trips to visit production facilities. The master plan was for 8,300 aircraft, but until the respective governments paid up, there would be nothing. The Feddens returned to the Mill in 1952. He retired again.

A name from the past resurfaced, Dowty, an engineer from the 1930s. Would Fedden be interested in being a consultant to the Dowty Group? He accepted right away. The Feddens were now based in the Queen's Hotel, Cheltenham, Gloucestershire. He immediately started travelling and representing Dowty throughout the world. He was still well known and had many contacts. He renewed old contacts in West Germany who had been adversaries during the war, Messerschmitt, Heinkel, and Dornier. He also discovered his old yacht in New Zealand. He retired again in 1960, aged seventy-five.

He was really proud of Cranfield and wanted it to remain the epitome of aeronautical engineering training and not be diluted by being associated with other institutions. By 1966, it had received a Royal Charter and was known as the Cranfield Institute of Technology, not quite what Fedden would have liked, but a satisfactory compromise. The Royal Aeronautical Society was involved in many discussions. The Society initiated a committee 'to make recommendations to the Government on general aviation policy'.

In his eighties, he devoted his efforts to protecting the banks of the Usk river and improving the fishing, His aerodynamic background guided him to install railway sleeper pads to cut down the erosion due to eddies. Weirs, banks, and pools were added in true Fedden fashion, fully documented. An engineer to the end, he surrounded himself with his created environment, which he could control – well, the trout in the river would decide which fly to bite, maybe not Sir Roy Fedden's every time!

He passed away in Wales on the 21 November 1973, age eighty-eight.

3

THE BRISTOL HERCULES PEDIGREE

I am indebted to the work of Alec Lumsden and his book *British Piston Aero-Engines and their Aircraft* published by Airlife Publishing Ltd in 1994, which was used as the basis for the statistics in this chapter.

The Hercules engine did not just happen. It was the result of a team, led by Fedden and Butler, that had developed and produced a long line of aero-engines, some more successful than others. The pedigree stretched back to the First World War when Fedden, working for Brazil Straker & Co., worked successfully on curing problems with the American Curtiss OX-5 aero-engine. The company also was the only production facility authorised to produce the Rolls-Royce Hawk and Falcon aero-engines under licence.

Prior to this, it was the French aero-engines that had cornered the market for supplying the power for aeroplane manufacturers. Gregoire, ENV, Gnome, Renault, Le Rhone, and Clerget engines were used by the aeroplane manufacturers, including Bristol. The Government slowly realised that it was not a good idea to be beholden to a foreign source to power their country's aeroplanes. Both the RFC and RNAS had become reliant on French aero-engines. The start of the First World War accelerated this thinking,

The Hercules: The Other Engine That Helped Win the War

as France and its factories would be vulnerable to possible attack and destruction. The race was on, but it took until 1916 for the first British engine to appear, the Bentley AR 1. In 1917, the Admiralty Air Department drew up a specification for a static, air-cooled, radial aero-engine producing more than 300 hp.

Fedden had obviously considered the different types of aero-engines that could be designed, such as air-cooled vs liquid-cooled and radial vs inline. The inline engine presented a smaller streamlined frontal area and seemed more appropriate for high boost, high altitude, and high-speed aircraft such as interceptors, fighters, and fast bombers. Its greatest vulnerability was damage to its cooling system by ground fire or an attacking fighter. Relative cooling efficiency is debatable. The rear cylinders on a radial engine are not cooled as well and the disturbed air adds to the total drag component. The liquid-cooled engine is more efficient because liquid is denser than air and can remove more heat. However, its nemesis can be the difficulty of getting uniform cooling despite the use of many thermostats. The air-cooled engine presents a larger frontal area, which can be a detriment to pilot visibility. It is, however, more rugged and can develop more power at lower altitudes. It lends itself to low altitude, naval ground attack and long endurance patrol aircraft.

The air-cooled engines, Armstrong-Siddeley, Bristol, Pratt & Whitney (US), and Wright (US), were single or twin-row radial, spokes of a wheel configuration engines. The liquid-cooled engines, Rolls-Royce, were in a V, Merlin, or X-shaped, Vulture, configuration. The two other major types of engine used were the poppet valve engine or the sleeve valve engine. Each engine would be given a distinctive Series or Mark number to differentiate it from similar engines. Fedden had obviously examined this challenge early on and decided on the air-cooled route, initially with the poppet-valve and later with the sleeve-valve engines.

Fedden and Butler immediately designed a fourteen-cylinder, helical configuration, air-cooled, radial engine for a competition set by Air Department at the Admiralty, as described above (see page 47). It was called the Brazil Straker Mercury. Frank

The Bristol Hercules Pedigree

Barnwell, British and Colonial Aeroplane Company, was looking for an engine to replace the Sunbeam Arab in the Scout Type F. The Cosmos Mercury engine was installed and flew in September 1918. In early 1919 it was tested at the Royal Aircraft Establishment and established records.

Fedden could see that more powerful engines would be required as the war continued, and when it ended, to meet the civilian aviation transport requirements. Brazil Straker's rival engines were the Siddeley-Deasy Jaguar and the Napier Lion (450 hp, 336 kw). Fedden decided that the company would build a nine-cylinder, air-cooled radial engine that would produce 500 hp (373 kw). He called it Jupiter. The company had no contract for this 28.7-litre (1,753 cu in) engine and built it on pure speculation, a bold and courageous move during wartime. It ran in October 1918 just before the cease of hostilities; but what did the future hold for Brazil Straker?

The company fragmented; Brazil himself left and the pieces were picked up by Cosmos, an Anglo-American financial group, who renamed it Cosmos Engineering Group, and basically told Fedden to carry on as usual. In addition, the government encouraged further development of the Jupiter engine. Barnwell was now interested in the Jupiter engine, and it was installed in the Bristol Badger. Fedden also continued work on a new engine, the Lucifer. The wartime contracts had dried up and it was a desperate time for manufacturing companies. Cosmos reached out to Sopwith for a seaplane installation and the Westland company. Then disaster struck; Cosmos Engineering went into liquidation. As mentioned earlier, the British and Colonial Aeroplane Company bought out the assets of Cosmos Engineering on 29 July 1920, Fedden included. This was fortuitous, as it turned out. This was the start of the pedigree which would eventually culminate in the Bristol Hercules and Centaurus engines. The demand for engines had ceased with the end of the war and the Jupiter, at that time, did not have any standing orders.

Fedden would spend the next twelve years successfully developing the poppet-valve engine with such names as the

Jupiter, Mercury, and Pegasus. He would then turn his attention to something he had pondered for twenty-eight years, the Argyll (120 hp, 104 kw) sleeve-valve engine. The Perseus and Aquila engines would gain the necessary experience with the sleeve-valve engine to enable the design, development, and production of the now famous Hercules engine.

The following table shows the Bristol aeroplanes that used their own engines in subsequent years up until the Hercules in 1936.

Bristol Engines Used in Bristol Aeroplanes 1920–1936

Year	Bristol Engine HP/KW*	Aeroplane	Type**
1920	Jupiter II (450 hp, 336 kw)	Bullet	D
1920	Jupiter III (435 hp, 324 kw)	Seely	CT
1921	Jupiter IV (425 hp, 317 kw)	Brandon/Ten Seater	CT
1922	Jupiter III/IV (425 hp, 317 kw)	Bullfinch	PD
1922	Lucifer 104 (140 hp, 104 kw)	M1D	PD
1922	Jupiter (510 hp, 380 kw)	Racer	PD
1923	Jupiter (510 hp, 380 kw)	Taxiplane/Trainer	CT/T
1923	Jupiter IV, VI, VIII (450 hp, 336 kw)	Bloodhound	F
1923	Jupiter IV, VI (450 hp, 336 kw)	Jupiter Fighter/Trainer	FT
1924	Cherub III (36 hp, 27 kw)	Brownie	PD
1925	Jupiter VI (450 hp, 336 kw)	Boarhound/Beaver	F/CT
1925	Jupiter VI (450 hp, 336 kw)	Type 92	PD
1926	Jupiter VI (525 h, 392 kw)	Badminton	PD
1927	Jupiter VI (450 hp, 336 kw) Twin	Bagshot	F/PD
1927	**Jupiter VIIF (490 hp, 365 kw)**	**Bulldog**	F
1927	Mercury II (485 hp, 362 kw)	Type 101	PD
1928	Mercury IIA (480 hp, 358 kw)	Type 107 Bullpup	F
1929	Jupiter XIF (490 hp, 365 kw)	Type 109	CT
1930	Neptune I (290 hp, 216 kw)	Type 110A	CT

The Bristol Hercules Pedigree

Year	Bristol Engine HP/KW*	Aeroplane	Type**
1931	Pegasus IM3 (650 hp, 485 kw)	Type 118/120	GP
1934	Mercury VIS2 (640 hp, 477 kw)	Type 123/133	F
1935	**Pegasus XXII (1,010 hp, 753 kw) Twin**	**Type 130 Bombay**	CT
1935	Mercury VIS2 (650 hp, 485 kw) Twin	Type 142	PD/CT
1936	Aquila I (500 hp, 373 kw) Twin	Type 143	P/CT
1936	Pegasus PEVIS (460 hp, 343 kw)	Type 138	PD
1936	Taurus II (1,050 hp, 783 kw)	Type 148/Beaufort	F/PD
1936	**Mercury VIII (840 hp, 626 kw) Twin**	**Blenheim/ Bolingbroke**	B/F

* Representative version and power rating
** B Bomber, CT Commercial/Transport, F Fighter, GP General purpose, T Trainer PD Prototype/Demonstrator

Bold: Successful aircraft in production numbers and service.

The Bristol engines were so popular that many other aeroplane manufacturers counted on using the engines in their aeroplanes. The following table shows this diversity.

Manufacturers that Used/Tested Bristol Engines 1920–1936

Poppet Valve Engine	Manufacturer
JUPITER	Airco, Blackburn, Boulton & Paul, De Havilland, Fairey, Fokker, Gloster, Gloucestershire, Handley Page, Hawker, Junkers, Parnall, Saro, Saunders, Short, Supermarine, Vickers, Westland
ORION	Gloster

The Hercules: The Other Engine That Helped Win the War

Poppet Valve Engine	Manufacturer
LUCIFER	AVRO, Handley Page, Parnall,
CHERUB	ANEC, Avro, BAC, Beardmore, Cranwell, De Bruyne-Maas, de Havilland, Granger, Halton, Hawker, Heath, Martin, Mignet, Messerschmitt, Parnall, RAE, Short, Supermarine, Vickers, Westland
TITAN	Avro
MERCURY	Airspeed, Blackburn, Boulton & Paul, Fairey, General Aircraft, Gloster, Hawker, Miles, Short-Bristow, Supermarine, Vickers, Westland
PEGASUS	Blackburn, Boulton & Paul, Douglas, Farey, Fokker, Handley Page, Hawker, Junkers, Koolhoven, Parnall, PZL, Saro, Short, Supermarine, Vickers, Westland
DRACO	Westland
PHOENIX	Westland
HYDRA	Hawker

Sleeve Valve Engine	Manufacturer
PERSEUS	Blackburn, De Havilland, Gloster, Hawker, Saro, Short, Vickers, Westland
AQUILA	Vickers
TAURUS	Fairey, Gloster

This is a good time to define some commonly used piston engine terms. The aero-engines were four-stroke, or four-cycle, engines and were devised by the German engineer Nikolaus Otto. The four strokes are known as the 'Otto Cycle'. These strokes are:

1) Intake: the air-fuel mixture is drawn into the cylinder
   a. Intake valve/intake sleeve valve ports　　　　　　OPEN
2) Compression: the air fuel mixture is compressed

      a. Intake/Exhaust valves/sleeve valve ports   CLOSED
3) Combustion: the air-fuel mixture is ignited producing the downward power stroke
      a. Intake/Exhaust valves/sleeve valve ports   CLOSED
4) Exhaust: spent air-fuel mixture expelled
      a. Exhaust valve/exhaust sleeve valve ports   OPEN

This completes the four cycles of the piston to produce power. The power produced by the engine is measured in two ways, simplistically:

1) Brake Horse Power (BHP) is the output power of the engine
2) Horse Power (HP) is the output power of a mechanical system (propeller) on the engine

Horsepower	Expressed in the Metric system as Kilowatts (kw or kW)
Engine Capacity	Total swept volume of the pistons inside all the cylinders
Un-supercharged	Normally aspirated
Supercharged	Process of getting more air-fuel mixture into the cylinder using pressure generated by blowers. These blowers could have different speed settings for different phases of flight. There could also be multi-stages.

Poppet-Valve Engine

Cylinder	Tubular section in which the piston works.
Bore	Diameter of cylinder
Piston	Capped tubular piece moving up and down vertically in cylinder
Stroke	Distance the piston moves up and down in the cylinder

Cylinder head	Caps top of cylinder, contains inlet and exhaust valves
Valve	Device in cylinder head that regulates flow in and out of cylinder
Camshaft	Rotating shaft with special protruding shapes, the cams
Rocker	Transmits the rotational movement of the camshaft cams to operate the valves in sequence

Sleeve-Valve Engine

Cylinder	Five port tubular section in which the piston and sleeve work
Bore	Diameter of cylinder
Piston	Capped tubular piece moving up and down vertically within sleeve in the cylinder
Stroke	Distance the piston moves up and down in the cylinder
Cylinder head	Caps top of cylinder junkhead
Sleeve	Tubular section which simultaneously moves in a vertical and horizontal direction causing an elliptical movement in the cylinder and opening and closing its four ports, Fourth port inlet and exhaust alternatively

The following is an abbreviated general list of the engines preceding the Hercules, which the curious reader may wish to research further. These engines were the building blocks of the foundation of knowledge, development, and production of the Hercules engine.

Poppet-Valve Engines

Mercury 1917

The Bristol Hercules Pedigree

Not to be confused with the later 1926 Bristol Mercury nine-cylinder engine.

Brazil Straker/Cosmos Mercury I 347 hp, 259 kw 14-cylinder, 20.01 L 1,223 cu in, air-cooled, double-row, helical, radial engine

Brazil Straker did get an order for 200 engines, but the Mercury did not enter production due the Armistice. The specification was awarded to the ABC Dragonfly.

Bristol 21A Scout F.1

Jupiter 1918
9-cylinder, 28.7 L 1,753 cu in air-cooled, single-row, radial engine

Fedden realised that the Mercury engine as a prototype was successful, but that for the future it would require a complete redesign and performance improvements. The Jupiter featured three carburettors, each one feeding three of the engine's nine cylinders via a spiral deflector housed inside the induction chamber. The Jupiter was otherwise fairly standard in design, but did feature four valves per cylinder, which was uncommon at the time. The cylinders were machined from steel forgings, and the cast cylinder heads were later replaced with aluminium alloy. In 1927, a change was made to move to a forged head design due to problems with the castings. The Jupiter was a very successful aero-engine, eventually being built by Bristol after 1920 and under licence across the world. It had over thirty variants and reputedly powered over eighty per cent of the aeroplanes exhibited at the 1929 Paris Air Show, from fighters, the British Bulldog, to airliners, the Handley Page HP 42, to flying boats, the twelve-engined Dornier Do X, the largest, heaviest, and most powerful flying boat in the world when it was produced. It certainly proved itself in Aerial Derbies, climb-to-altitude records, and long-distance flights. In 1926, a Jupiter-engined Bristol Bloodhound completed an endurance test of 225 hours without any failures. The engine matured into one of the most reliable on the market. It was the first air-cooled engine to pass the Air Ministry full-throttle test, the first to be equipped with automatic boost control, and the first to be fitted to airliners.

Brazil Straker Jupiter I, 400 hp, 298 kw
Cosmos Jupiter II, 400 hp, 298 kw
Bristol 23A Badger II
Bristol 32 Bullet
Sopwith Schneider
Westland Limousine II
Bristol Jupiter II 1923 400 hp, 298 kw
Boulton & Paul P25 Bugle I
Bristol 32A Bullet
Westland Weasel

Jupiter Series

Un-supercharged: II, III, IV, V, VI, VIII, IX, XI
Supercharged: VII, X variants

Series	Year	HP/KW	Aeroplane Example
III	1923	400/298	Bristol 52 MFA Bullfinch I
V	1925	480/358	Bristol 84 Bloodhound
VIA	1927	440/328	Vickers 143 Bolivian Scout
VIFH	1932	440/328	Bristol 105A Bulldog IIA
VIFL	1932	440/328	
VIFM	1932	440/328	
VIFS	1932	440/328	
VII	1928	375/280	Junkers W34
VIIF	1929	375/280	Handley Page HP33 Hinaidi I
VIIF.P	1929	375/280	Gloster Gamecock (special)
VIIIF	1929	460/343	Boulton & Paul P29 Sidestrand III
VIIF.P	1929	460/343	Westland Wapiti I, IA, II, IIA
IX		525/392	Supermarine Seagull II

The Bristol Hercules Pedigree

Series	Year	HP/KW	Aeroplane Example
X		470/351	Short S10 Gurnard I
XF	1931	540/403	Blackburn BT1 Beagle
XFA		483/360	Bristol 118
XFAM		580/433	Hawker Hart
XFBM		580/433	Vickers 214 Vildebeest IV
XFS	1931	580/433	De Havilland DH72
XI		460/343	Fokker F VIIA
XIF		500/373	Short S8 Calcutta
XIF.P	1932	525/392	Saro A7 Severn

Jupiter IV 1926 430 hp, 321 kw

Boulton & Paul P.25 Bugle I	Fairey Flycatcher I
Bristol 52 MFA Bullfinch I	Gloucestershire Mars VI Nighthawk
Bristol 53 MFB Bullfinch II	Gloucestershire Grebe I
Bristol 72 Racer	Gloucestershire Gamecock I
Bristol 72 Racer special Jupiter IV racing engine	Handley Page HP12 O/400
Bristol 75 Ten-seater	Hawker Duiker 7/22
Bristol 75A Express Freighter	Hawker Hedgehog 37/22
Bristol 76 Jupiter Fighter	Hawker Woodcock II 25/22
Bristol 79 Brandon	Parnall Plover
Bristol 84, 84A, 84B Bloodhound	Short S3 Springbok I, II
Bristol 89, 89A	Short S3B Chamois
Bristol 93, 93B Boarhound I, II, 93A Beaver	Vickers FB 27 Vimy
De Havilland DH 42 Dormouse	Vickers 113 Vespa I
De Havilland DH 42B Dingo II	Vickers 144 Vimy Trainer
De Havilland DH 50	

Jupiter VI 1927 480/520 hp, 358/388 kw

The Hercules: The Other Engine That Helped Win the War

Airco DH 9 'M'pala I	Gloster Gamecock II
Airco DH 9AJ Stag	Gloster Goring
Boulton & Paul P 29 Sidestrand I	Gloster Gambet
Bristol 84 Bloodhound normal & derated	Handley Page HP27 W9a Hampstead
Bristol 89, 89A normal & derated	Hawker Hawfinch
Bristol 92 Laboratory	Hawker Heron
Bristol 93A Beaver	Junkers F13
Bristol 93B Boarhound II	Saunders A4 Medina
Bristol 95 Bagshot	Short S6 Sturgeon I
Bristol 99, 99A Badminton	Vickers 113 Vespa I
Bristol 99A Badminton short stroke racing engine	Vickers 119 Vespa II
Bristol 101	Vickers 121 Wibault Scout
Bristol 107 Bullpup	Vickers 131 Valiant
Bristol 124 Bulldog TM	Vickers 149 Vespa III
De Havilland DH 50	Vickers 159 Vimy Trainer
De Havilland DH 61 Giant Moth	Westland Wapiti prototype
De Havilland DH 66 Hercules	Westland Wapiti I
De Havilland DH 67 Survey	Westland Westbury
Fairey Ferret I, II	Westland Witch I

Jupiter VIII 1929 440 hp, 328 kw

Airco DH 9 'M'pala II	Handley Page HP 34 Hare Type H
Blackburn T 5 Ripon IIF	Hawker Harrier
Boulton & Paul P 29 Sidestrand	Hawker Hart
Bristol 84 Bloodhound	Vickers 132 Vildebeest I
Bristol 109	Vickers 150B 19/27
De Havilland DH 67B Survey	Westland Wapiti I, IA, II, IIA
Fairey IIIF	Westland Westbury
Handley Page HP 33 Hinaidi prototype I, II Type M	Westland Witch

Lucifer 1919
3 cylinder, 8.0 L 487 cu in, air cooled, single-row, radial engine
Cosmos

Lucifer I 1919 80/100 hp
Avro 504K Lucifer
Boulton & Paul P.10
Bristol

Lucifer II 122 hp, 91 kw
Avro 504N
Bristol 73 Taxiplane
Bristol 83 School Machine

Lucifer III 1923 118 hp, 88 kw
Bristol 73 Taxiplane
Bristol 83 School Machine
Parnall Peto

Lucifer IV 1925 140 hp, 104 kw
Avro 504N
Bristol 77 M.1D special engine
Bristol 83 School Machine
Handley Page HP 32 Hamlet Type D
Parnall Peto

Cherub 2-cylinder, 1,095 cc, 67 cu in, air-cooled, horizontally-opposed engine

Series	Year	HP/KW	Aeroplane Example
I	1923	32/24	Avro 562 Avis
II	1924	34/25	Short S4 satellite

Cherub III 1924 36 hp, 27 kw

Anec II	Heath Parasol
BAC Drone	Martin Monoplane

The Hercules: The Other Engine That Helped Win the War

Beardmore WB XXIV Wee Bee	Mignet HM 14 Pou de Ciel
Bristol 91A Brownie I	Parnall Pixie III, IIIA
Bristol 91B Brownie II	RAE Hurricane
Bristol 98 Brownie III	RAE PB Scarab
Cranwell CLA 4, 4A	Supermarine Sparrow II
De Bruyne-Maas Ladybird	Vickers 98 Vagabond
De Havilland DH 53 Humming Bird	Westland Woodpigeon I
Hawker HAC 1 Mayfly	Westland-Hill Pterodactyl IA
Hawker Cygnet	

Orion 1926
9-cylinder 28.7 L 1,753 cu in single-row, air-cooled, radial engine
It was based on the Jupiter III with exhaust turbo-charger, abandoned due metallurgy concerns

Mercury 1926
9-cylinder, 24.9 L 1,520 cu in, single-row, air-cooled, radial engine

The Mercury engine was designed by Roy Fedden to replace the Jupiter, which was at the end of its useful life. Some of the Jupiter parts were used, the block for example, and the smaller Mercury began to achieve ever increasing horsepower. The Mercury XV used 100 octane fuel from the US, which allowed it to run at higher compression ratios and supercharger boost pressures. The Mercury was also the first British engine to be approved to have a variable-pitch propeller. A total of thirty Mercury variants were built using the Jupiter block and some variants shortened its stroke to increase the rpm for fast climbing fighter aircraft. Later variants returned to the longer stroke and similar piston speed, allowing gearing suitable for the long-range bomber. There were four variants that were un-supercharged. It was also the first, an important milestone, British engine to be approved to have a variable-pitch propeller. The Mercury would evolve into another successful engine, the Bristol Pegasus.

The Bristol Hercules Pedigree

Series	Year	HP/KW	Aeroplane Example
I	1926	808/603	Short-Bristow Crusader (7/26)*
II	1928	420/313	Bristol 101
IIA^	1928	440/328	Gloster SS18 (F10/27)
III	1929	484/362	Westland Interceptor (F10/27)
IIIA	1930	485/362	Westland C.O.W. Gun (F29/27)
IV	1929	485/362	Hawker Audax
IVA	1931	510/380	Bristol 105A Bulldog IIIA
IVS 2	1932	510/380	Bristol 105A Bulldog IV
SSE		390/291	Bristol 107 Bullpup
V		546/407	Bristol 118A
VIS	1933	605/451	Gloster Gauntlet I, II
VISP	1931	605/451	Hawker Fury (Persia) (F13/30)
VIS 2	1933	605/451	Bristol 142 'Britain First'**
VIA	1928	575/429	Hawker Hart
VIIA		560/418	Fokker G1
VIII	1935	825/615	Bristol 149 Blenheim V**
VIIIA	1934	840/630	Gloster Gladiator I (F7/30)
VIIIA		535/399	renamed Pegasus IU 2P
IX	1935	825/615	Blackburn B24 Skua I

Post 1936: X, XI, XII, XV, XVI, XX, 25, 26, 30, 31

* (7/26) The Government competition specification number

** In early 1934, Lord Rothermere, owner of the *Daily Mail* newspaper, challenged the British aviation industry to build 'the fastest commercial aeroplane in Europe, if not the world'. Bristol responded with the Blenheim, which went on to become a light bomber, night fighter, and long-range reconnaissance aeroplane.

The following symbols are used for both the Mercury and Pegasus engines.

F	Engine-speed induction fan
U	Induction fan geared to 500 ft, 152 m
L	Low induction fan geared to 1,500 ft, 457 m
M	Fan geared for moderate supercharge 5-7,000 ft: 1,524-2,133 m
P	Persia. Ran on local DTD 134 fuel. (Pusher, Pegasus)

The Hercules: The Other Engine That Helped Win the War

S   Fan geared for fully supercharge 10-15,000 ft: 3,048-4,572 m

2, 2A, 3, 4 Different airscrew reduction gear ratios

Titan 1928
5-cylinder 14.5 L 844 cu in, single-row, air-cooled, radial engine

Titan I 1928 205 hp, 153 kw
Avro 504N
Bristol 83 trainer
Bristol 110A

Titan IV 1928 205 hp, 153 kw
Bristol 83E Trainer

Phoenix 1928
9-cylinder 28.7L 1,753 cu in, single-row, air-cooled diesel engine

Phoenix I 1928 470 hp 351 kw
Westland Wapiti I

Phoenix IIM
Westland Wapiti I (J9102). Established world altitude record for diesel engines. 27,453 ft, on 11 May 1932

Neptune 1930
7-cylinder, 19.4 L 1,182 cu in single row, air-cooled, radial engine
7 cylinder

Neptune I 1930 290 hp, 216 kw
Bristol 110A

Bristol Hydra 1931
16-cylinder, 25.7 L 1,570 cu in double-row, air-cooled, radial engine
Hawker Harrier

Pegasus 1932
9-cylinder 28.7 L 1,753 cu in single-row, air-cooled, radial engine

The Pegasus engine was a natural evolution from the experience gained by the previous Jupiter and Mercury engines. The

The Bristol Hercules Pedigree

engine reverted to a long stroke version of the Mercury, similar to the Jupiter internal dimensions. Both engines displayed an improved power-to-weight ratio. The Pegasus laid claim to 'one pound-per-horsepower' and it achieved a higher maximum engine speed, which resulted in more horsepower. Gear-driven superchargers were introduced in production VIIs. In collaboration with RAE, experimental exhaust driven superchargers were fitted to the Jupiter III and IV. The engine established a height record in 1932 of 13,404m (43,976ft), the first flight over Mt. Everest, and in 1938 set a world long distance record. Approximately 32,000 engines were built.

Un-supercharged: IF, IU, HU series
Supercharged: II to 48 series

Series	Year	HP/KW	Aeroplane Example
IF	1932	546/407	
IM	1932	590/440	
IM.2	1933	590/440	Westland Wapiti
IM.3	1933	590/440	Boulton & Paul P75 Overstrand I
IS.2		546/407	
IS.3	1933	550/410	Vickers 250 Vespa VII
IU.2	1928	575/429	Hawker Hart
IU.2P		535/399	
IU.3		535/399	
IU.3P		535/399	
IIL.2	1934	625/466	Gloster Goring
IIL.2P	1934	625/466	Supermarine Seagull V
IIL.3	1934	625/466	Vickers 252 Vildebeest XI
IIM	1934	620/462	Fairey TSR I
IIM.2	1934	620/462	Hawker Osprey
IIM.2P			Supermarine Walrus I
IIM.3	1934	620/462	Blackburn B5 Baffin
IIU		595/444	
LC		595/444	

The Hercules: The Other Engine That Helped Win the War

Series	Year	HP/KW	Aeroplane Example
MC		595/444	
IIIM.2	1935	690/515	Fairey Swordfish I
IIIS.2	1935	700/522	
IV	1933	700/522	Koolhoven FK52
V	1935	690/515	
PE.6S		500/373	Bristol 138A
VIP		690/515	Supermarine Walrus I, II
VII		620/462	
VIII		700/522	
IX	1934	725/541	Blackburn T9B Shark III
X	1936	980/731	Bristol 130 Bombay prototype
			Saro A27 London II
			Supermarine Stranraer
			Vickers 281 Wellesley
			Vickers 285 Wellington prototype
XC	1936	900/671	Short S23, S33 Empire
XI	1936	850/634	
XII	1936	850/634	
XIII		960/716	
XV		725/541	Handley Page HP52 Hampden prototype

Pegasus IIIM3 1935 690 hp, 515 kw

Bristol Blackburn T9A Shark II	Hawker PV4 (G4/31)
Bristol Blackburn T9B Shark III	Parnall (G4/31)
Bristol 130 Bombay prototype (C26/31)	Saro A27 London I
Fairey Seal	Supermarine Stranraer
Fairey TSR II	Vickers 212 Vellox
Fairey Swordfish I	Vickers 246 Wellesley prototype (G4/31)
Handley Page HP47	Vickers 266 Vincent

Handley Page HP51	Westland PV7 Wallace prototype (G4/31)
Hawker Hart	

Post 1936: XVII, XVIII, XIX, XX, 21, 22, 23, 24, 25, 26, 26, 27, 29, 30, 32, 38, 48

Draco 1935
9-cylinder, 28.7 L 1,753 cu in, fuel-injected, single-row, air-cooled radial engine
Westland Wapiti

Sleeve-Valve Engines

The seeds of the Bristol Hercules engine were sown in 1914 when a young Roy Fedden, then Chief Engineer and Technical Director of Straker-Squire, discovered and was impressed by the Argyll six-cylinder, water-cooled aero-engine. A radical departure from the normal camshaft, pushrod, and valve engine, it used an ingenious single-sleeve valve system to achieve the same four strokes of the internal combustion engine. The system had been patented in 1909 by Scotsman Peter Burt and Canadian James McCollum as part of Argyll Motors Ltd.

In 1911, Argylls, the new company name, exhibited a four-stroke, sleeve-valve engine at the Motor Show. This triggered a case against Argylls for infringement of the Knight and Kilbourne Co. 1905 original patent. It was dismissed on the grounds that an engine built to that patent came nowhere near the Argylls' rated horsepower and that Argylls featured other technical features, and therefore could not be the same engine. The Burt engine was an open sleeve type, driven from the crankshaft side, while the McCollum engine had a sleeve in the cylinder head and a more complex port arrangement. The input of both inventors resulted in a sleeve valve that was given both vertical and partial rotary motion to open and close ports.

Fedden grasped the concept immediately and remembered it, and this knowledge would become the foundation of his sleeve-valve engines many years later.

Perseus 1932
9-cylinder, 24.9 L (1,520 cu in), single-row, air cooled, radial engine

In 1926, Harry Ricardo of the Royal Aircraft Establishment published a series of papers on the principles of the sleeve-valve engine. This engine had better volumetric efficiency, ability to operate at higher rotational speeds, better fuel efficiency, and more compact design than the poppet-valve engine. Development problems took time to solve and by 1932 the Perseus, a Mercury-sized engine, became the first sleeve-valve engine to be put into full-scale production. The Perseus was used mainly in military applications and civilian flying boats. The engine's claim to fame was that it enabled the development of the twin-row sleeve-valve radial engines to follow, notably the Hercules and Centaurus.

Series	Year	HP/KW	Aeroplane Example
I	1932	515	Prototype engine
I.A	1932	515	Bristol 105A Bulldog IVA
II-L	1932	665	Short L17 Scylla
III	1932	665	Hawker Hart
VIII	1936	745	Vickers 286 Vildebeest IV

Post 1936: X, XA, XI, XII, XIIC, XII-CI, XIV-C, XVI, XVIC

Aquila 1934
9-cylinder, 15.6 L (950 cu in), single-row, air-cooled, radial engine

The Bristol Hercules Pedigree

The Aquila was a smaller version of the Perseus which developed close to the power of the later Jupiter engines, but by that time the focus was on larger, more powerful engines. It never went into full scale production.

Taurus 1936 (First run)
14-cylinder, 25.4 L (1,550 cu in), twin-row, air-cooled, radial engine

Did not really develop until after the Hercules engine in 1936. With the experience gained from the Perseus and Aquila engines, the next design step was to meet the expected wartime horsepower demands by creating a larger twin-row, sleeve-valve engine. The foundation of knowledge was there, but could Fedden take this knowledge to the next level of power and performance? He did, and it was called the Bristol Hercules.

4

BRISTOL HERCULES: THE NUTS AND BOLTS

Roy Fedden had not forgotten about the Burt-McCollum type of mono-sleeve valve engine. A small crank oscillates a single sleeve in a simple cycloidal way with a rotary as well as an up-and-down motion. This design increased heat dissipation, reduced the number and type of parts, improved 'valve' (ports) timing, and created a swirling motion of the combustible mixture. The resulting mixture now allowed higher compression ratios, which resulted in more power and economy.

With the Mercury engine in production, he simultaneously worked on the Pegasus poppet-valve engine and the Perseus sleeve-valve engine. In 1925 he was reminded of the sleeve-valve engine by that series of papers by Harry Ricardo. They extolled the virtues of the sleeve valve design over the poppet-valve engine. It had better volumetric efficiency and the ability to operate at higher rpm, with the result that a smaller compact engine could produce the same power as a larger poppet-valve engine.

In 1926 Fedden decided to adopt the mono-sleeve design. It took seven years of trial and error to get his first sleeve-valve engine working, Bristol had to absorb great expense during the experimental stage. The basic design which he adopted had a four-port sleeve, two inlet and two exhaust, one port acted as both inlet and exhaust. This matched a five-port cylinder barrel, three for inlet and two for exhaust. He created nine, fourteen, and eighteen-cylinder sleeve-valve engines, which were light and powerful. The greatest of these during the late 1930s and wartime would be the fourteen-cylinder Bristol Hercules engine.

Hercules 1936

14-cylinder, 38.7 L (2,360 cu in), two-row, air-cooled, radial engine

The sleeve-valve was a thin-walled tube which slid within the cylinder between the piston and the cylinder wall. The sleeve had apertures cut in it. The function of the sleeve was to act as inlet and exhaust valves, at the correct point in each stroke of the four-stroke cycle, when the apertures were opposite ports in the cylinder head. The sleeve was attached to a crank which moved it vertically and rotated it horizontally to allow it to close or open the necessary apertures and ports in sequence. The challenge of design, manufacture, assembly, and maintenance of the sleeve-valve mechanism was just about beyond comprehension, except for Fedden.

Fedden and Butler, after extensive study, decided to build a twelve-cylinder inverted V sleeve-valve engine. The initial weak link in the sleeve-valve engine was fairly evident, the sleeve. Any distortion of the sleeve would affect the efficiency of the cylinder and piston lubrication. A variety of metal was tried, nickel-cast-iron and steel valves, to no avail. Next was stainless steel, the soft sleeve failed but the hard 'nitride' surface sleeve allowed the research to continue. By 1927, Fedden and Harvey Mansell had run a two-cylinder V-shaped, air-cooled sleeve-valve engine, and in 1931 got a one-cylinder sleeve-valve engine to run for 100 hours. Based on this, Fedden decided that his first engine would be a nine-cylinder, sleeve-valve, radial engine using the previous Mercury-sized cylinders. He called it the Perseus and it first ran in 1932 and flew in a Bristol Bulldog in 1933.

In 1935, Fedden commenced work on the Hercules using fourteen Perseus cylinders in two staggered rows. All the sleeves were operated by an impressive (an understatement) series of gears situated on the front of the engine. By January 1936 the bench tests were showing promising results, and by November the Hercules had been sufficiently developed to be exhibited at the Paris Air Salon, where it stole the show as the most powerful type-tested engine in the world. In 1937, it was test flown in a Northrop Gamma. The Hercules went on to become one of the most successful piston engines ever built. Over 65,000 engines powered civil and military aircraft for the next three decades.

A description of the Hercules engine components will be followed by a listing of the engines in the series, and then a listing of the manufacturers that used the engine in their aircraft.

Hercules VI, XI, XVI, XVII, XVIII Engine

These engines employ the two-speed supercharger, down-draught carburettor, dual ignition from two magnetos and screened high-tension distributing harness. The propeller shaft, which is driven through a reduction gear, will take various types of hydraulic and electric constant-speed propellers. Accessories, others than those directly associated with the engine, are mounted on a gearbox which is an entirely separate mechanism attached to the bulkhead and driven by a cardan shaft from the rear cover.

Although all the engines are similar, there are certain differences between types brought about either by development or particular operational requirements, or both. Some engines have a fully automatic carburettor, which dispenses with a hand-operated mixture control, whilst others combine this feature with a lower rate of supercharge. For this reason, it was very important that the aircrew be fully aware of the Engine Reference Data and the Operating & Performance Data of their particular aircraft.

CRANKCASE: is in three sections, the front, centre, and rear. The sections are joined at the respective centre lines of the rows of cylinders and held together by seven large bolts. The rear set of bolts extend rearward to secure the supercharger and engine mounting ring.

SLEEVE-DRIVING MECHANISM: the front section of the crankcase holds the sleeve-valve operating mechanism. It has fourteen cranks, one for each sleeve. The rear cylinder cranks have long spindles passing between the front connecting rods to the rear. The short front cranks rotate in roller bearings housed in the front section. The rear cranks are housed in the centre section. The front end of all sleeve cranks are force-lubricated plain bearings located in the front cover.

All cranks are driven by a gear on the crankshaft which meshes with seven gears ringed about it. These gears run on needle bearings on the layshafts extending from the front crankcase to the front cover. Attached to each layshaft gear is a pinion meshing with two sleeve-driving gears, one to a front and one to a rear sleeve crank, rotating them at half engine speed.

FRONT COVER: has a small oil drain pump at the bottom driven by the No. 8 sleeve crank. The propeller constant-speed unit is mounted externally on the upper front cover. It is driven by the No. 2 sleeve crank. A ring of studs provides attachment points for the exhaust manifold.

CRANKSHAFT: is a two-throw design with three parts, front crankweb, rear crankweb, and centre section. The centre section has two crankpins which are attached to the front and rear crankwebs by split type maneton joints. The front joint has two bolts and the rear one has one bolt. The complete crankshaft is carried in three roller bearings, one to each section, housed in the crankcase walls. In addition, the front crankweb is supported by a roller bearing in the front cover. Balance weights are fitted to the front and rear crankwebs. In later engines 'Salomon dampers',

The Hercules: The Other Engine That Helped Win the War

floating spherical weights, are also incorporated within the balance weights to reduce vibrations.

CONNECTING-RODS: the two connecting-rod assemblies are identical. Each comprises a master rod to which six articulated rods are attached by wrist pins. Oil tightness is achieved on the big-end bush or bearing by fixed or floating oil retainers, depending on whether it is a bush or bearing type. Engines with floating type big-end bearings have the fixed oil retainers attached to the master rods while the fixed type are secured to the crankshaft.

CYLINDER: the fourteen cylinders are arranged in two rows of seven. The master rods are in cylinder No. 4 and No. 11. The oil sump is located behind No. 8 cylinder. Each cylinder has an aluminium drop forged detachable head called a junkhead. It projects into the sleeve bore and seals, by way of piston-type seals, the combustion chamber. There are two spark plugs fitted in the base of the head. The head is finned for heat-radiating purposes.

About one third the way down the cylinder wall there are five ports, the two in the front are exhaust, the two in the side and the one in the rear being inlet. An induction belt connects the three inlet ports.

SLEEVE: an open cylinder with four ports cut in its top end. It is joined at the bottom to a crank via a ball-joint, so that the sleeve follows the movement of the crank.

PISTONS: the aluminium-alloy pistons are attached to the connecting rods by hollow case-hardened gudgeon-pins. Twelve connecting-rods are identical with the two master connecting-rods having a slightly larger diameter to allow for additional lubrication.

SLEEVE VALVE OPERATION: the sleeve slides up and down in the cylinder and at the same time is given a rotary oscillatory motion. The movement of any point on the sleeve is elliptical. The four ports are designated, two inlet, two exhaust, and one combined use inlet and exhaust. During the compression and expansion stroke the ports are traversing at the top of their elliptical arc and are above the bottom of the junkhead. Thus, they are shielded from the effects of combustion while the cylinder ports are closed off by the sleeve wall. During the exhaust and induction stroke the ports are traversing at the lower level of their elliptical arc, first uncovering the exhaust ports and then the inlet ports. The dual purpose sleeve port moving back and forth from exhaust to inlet port.

SUPERCHARGER: the two-speed supercharger is mounted on the rear of the crankcase by extended bolts. It has a double-shrouded aluminium-alloy impeller which takes carburettor mixture and passes it on centrifugally through seven outlets to an induction pipe feeding front and rear cylinders. A tailshaft, passing through the hollow impeller shaft, drives the supercharger, via a spring-drive gear, and the rear cover gears. The starter jaw is mounted to the tailshaft.

The spring-drive gear provides two gear ratios to drive the three two-speed clutches. The springs dampen torsional variations from the crankshaft. The two-speed clutches in turn drive a centrifugally operated clutch attached to the impeller-shaft. Thus, the shaft is driven by friction between the centrifugal clutch-blocks and their housing. The three clutch units are spaced around the impeller-shaft to avoid any journal load to the high-speed impeller-shaft bearings. The clutches are operated by pistons actuated by oil pressure and controlled by a valve which selects the respective piston for high or low gear. A pair of revolving 'centrifugers', oil-cleaners, is fitted to the lowest clutch to enable easy maintenance.

REAR COVER: is mounted on the rear face of the supercharger and incorporates oil and fuel pumps, two magnetos, and rpm indicator. The magneto drive incorporates a spiral spline device for varying the ignition timing in relation to throttle position. It is connected to the throttle lever by a cam mechanism so as to give an advance setting while in the slow running position and a retard setting while in the full throttle position. An ATD coupling is fitted to prevent engine kick back when the throttle is set slightly open for start. The ATD retards the ignition during start by ten degrees from the setting determined by the cam and spiral drives and retards the ignition below 1,200 rpm. Above this speed, the ignition setting is by throttle position alone.

ENGINE CONTROLS: installation is simplified by connecting all main engine controls to a central set of levers all mounted on a shaft attached by brackets to the tear cover.

REDUCTION GEAR: the self-centralising epicyclic bevel gear type reduction gear is fitted to all Bristol produced engines (1945). It is attached to the front cover and driven from the crankshaft by a driving member. The propeller shaft is supported at the front end by a ball bearing and at the rear by a bush in the crankshaft bore. The reduction gear reduces the propeller speed to 0.444 crankshaft speed. The pressure oil for operating the propeller is transferred internally to the propeller hub.

EXHAUST SYSTEM: the exhaust ring is an annular chamber forming the leading edge of the engine cowling. It is attached by three tripods with one leg of the tripod attached to the reduction gear and the other two legs to the front cover. The apexes are secured to flexible mountings inside the ring. There are fourteen exhaust branch pipes between the cylinders and the ring, each serving a front and rear cylinder. The ring has one tail pipe outlet.

CONTROLLABLE GILLS: form a controllable cowl in the rear extension of the engine cowl. The gills control the air-flow over the engine for varying conditions of flight or ground running.

Bristol Hercules: The Nuts and Bolts

LUBRICATION: the oil pump is behind the rear cover at the port end of the transverse shaft, the other end of the shaft drives the fuel pump, which shares a common drive but in separate casings. The oil pump comprises the pressure pump and the larger scavenge pump ensuring oil is always available to the pressure side.

The pressure pump receives its oil from an internal channel via a union on the rear cover connecting the oil tank. The oil is delivered to the engine through the end of the hollow driven wheel spindle to which the oil is fed from ports in the pump casing. Oil passes under pressure to the main bearings from the chamber in the rear cover. The pressure is normally controlled by the relief valve, which bypasses oil back to the pump inlet.

A restrictor is provided in the relief valve for starting with cold thick oil. It allows the pressure to build up to 200 lb/sq in before the ball safety valve opens and bypasses the excess oil back to the inlet side of the scavenge pump. This high initial oil pressure device provides adequate start-up lubrication and allows immediate engine operation at temperatures as low as +5°C (41°F).

The main oil feed is routed from the pump up through passages in the rear cover casting to the tail-shaft bush, to a tube inside the tail-shaft, then to the crankshaft. This supply also lubricates all the bushes in the rear cover except the pump-drive shaft and the ignition control lever. Holes in the bore of the shaft feed the pump-drive shaft bushes and the ignition-control shaft bush is splash-lubricated, as are all the rear cover gears.

Two external oil connections lead from the rear cover:

1) Direct from the pump chamber to the sleeve-driving mechanism in the front with a branch line to the supercharger
2) Direct from the main feed system to the boost and mixture controls in the carburettor at the rear.

Two slots in the tail-shaft bush provide an intermittent oil feed through a small hole to supply a small quantity of oil first to the impeller-shaft bore and then to the bearings, gears, and centrifugal clutch. The two-speed clutches are lubricated by their operating oil.

Oil passes through the hollow crankpin and drilled passages in the three sections of the big-end bushes and master-rod bores. The oil retainers restrict leakage from the big-ends, whilst the floating retainers convey oil through the spring-loaded plungers to the wrist pins and their bushes. In both front and rear crankwebs an oil jet is fitted, which is connected to the main oil passage. Each contains a spring-loaded ball-valve, which opens at a pre-determined pressure to lubricate pistons and sleeves.

In the front cover, oil travels around an oil ring, flows through passages to the front ends of the sleeve cranks, and through the bores of the cranks to the sleeve ball-joints. The gears of the sleeve-diving mechanism are lubricated by splash oil. Oil from this ring also flows to the propeller governor unit.

Oil from the main supply in the crankshaft flows through the bore of the propeller shaft to the reduction-gear bevel-pinion bushes. The gears are lubricated by splash oil. Some propellers are operated by oil pressure, others are not.

The pistons, cylinders, and sleeves are all lubricated by splash oil supplemented by the spray from the crankshaft jets, The amount of oil distributed over the sleeve outer surfaces is regulated by a contracting ring around each sleeve, fitted near the lower end of the cylinder bore.

All drain oil flows to the sump, it is assisted by the front cover drain pump. From the sump, oil is drawn by the scavenge pump and passed through a jacket around the carburettor to warm the chokes and throttles before being transferred to the oil tank. The jacket has a cold oil relief valve.

CARBURETTOR: is the Hobson down-draught carburettor mounted on top of the supercharger. It is the duplex type, two carburettors in one, with each half having a float chamber, with its mechanism and fuel feed, a detachable choke and a delivery tube, diffuser, slow-running jet, and main jet. Three jets are installed: one power, one enrichment, and one supplementary, each serving both halves of the carburettor.

The carburettor is compensated for boost and altitude by means of two barometric capsule controls:

1) The boost control is connected to the induction pressure, which it regulates for each power condition until rated altitude is reached. It does this by progressively opening the throttle butterflies through the medium of an oil-operated servo control.
2) The mixture control is open to atmospheric pressure and as this falls the control adjusts the mixture strength.

On early engines the mixture strength had three manual selections, normal, automatic, or weak. Later engines had single lever carburettors which were fully automatic and provided the best mixture strength when the throttle was within the cruising range. An accelerator pump was incorporated to ensure rapid and smooth acceleration.

IGNITION: these series engines could be fitted with Simms, Rotax, or B.T.H. magnetos. They could be fitted to the port or starboard side of the engine and were driven by the variable drive and the A.T.D. coupling at one and a quarter crankshaft speed. The B.T.H. and Rotax magnetos have a fourteen-lobe cam with a single contact breaker. The Simms magneto has a seven-lobe cam and two contact breakers. The right contact breaker supplies the front cylinders while the left contact breaker supplies the rear cylinders. The contour of the cam in all three magnetos is designed to give the compensated firing necessitated by the articulation of the connecting rods which is a feature of the radial engine.

For starting purposes, a high-tension booster coil is connected to the magneto. Distribution of the current is by a starting (or trailing) electrode incorporated in the distributor rotor. The whole ignition system is screened and bonded to the engine structure to prevent radio interference. The H.T. cables are taken through flexible conduit to a circular metal casing attached to the rear casing. Then individual cables in flexible conduits are connected to the individual spark plugs inside a plug screening tube.

The fourteen cylinders are numbered as follows looking at the front of the engine:

1) Rear row of cylinders are the odd numbered cylinders. No. 1 is at the 12 'o' clock position and they are numbered sequentially clockwise as follows: 1; 3; 5; 7; 9; 11; 13.
2) Front row of cylinders are the even numbered cylinders. No. 8 is at the six 'o' clock position and they are numbered sequentially clockwise as follows: 8; 10; 12; 14; 2; 4; 6.

OPERATION

GROUND RUNNING BEFORE STARTING: standard control settings

  i. Turn on fuel cocks and close suction and pressure balance cocks.
  ii. Set the propeller speed control lever to the maximum rpm position and, with a Rotol electric propeller, check that
      a. Selector switch in AUTOMATIC
      b. Feather switch in NORMAL
      c. Master switch ON
  iii. If Extractor controls are fitted, one stroke for priming purposes is necessary.
  iv. Set the throttle lever just off the closed position. Caution: if opened too wide a backfire may occur when the engine starts. Never pump the throttle lever.
  v. Set the mixture control lever at NORMAL (series VI and XI). Later engines are not fitted with a hand operated mixture control.

vi. Set air intake heat control to COLD AIR.
vii. Open cowl gills fully unless the temperature is below 0°C (32°F).
viii. Set the supercharger lever to 'M" (low) setting.
ix. Close oil cooler shutters manually, if installed.

STARTING: make sure all ignition switches are turned off and turn the propeller by hand through two complete revolutions. Any resistance would indicate that there is fluid in the lower cylinders. To remove the fluid the spark plugs and induction pipes should be removed before rotating the propeller through several revolutions. Replace plugs and lockwire where necessary.

Set the cocks in the engine doping system to their correct position. Use high-volatility fuel as required by outside temperature. Prime pipes by the feel of the plunger. Switch on main magnetos and operate the booster coil or starter magneto. Press the starter button and while the engine is turning operate the doper pump vigorously at a rate according to the outside temperature, the lower the temperature the higher the rate. On some installations the booster coil operates when the starter is operated. The engine should start immediately.

If the engine fails to start, wait thirty seconds before further doping. After four attempts further investigation is required. Under cold conditions it may be necessary to continue doping after the engine has fired until it picks up on the carburettor. As soon as the engine is running regularly, switch off the booster coil (or starter magneto), turn off the doper cock and screw the pump plunger firmly home.

WARMING-UP: The oil pressure should rise immediately after starting. If not, shut down and investigate. The oil pressure should rise to 150-200 lb/sq in on starting with cold oil and slowly drop back to normal as the temperature increases. If pressure settles 10 lb/sq in below normal, or fluctuates, shut down engine for investigation. During very cold weather the engine should not be shut down, unless for obvious reasons, until normal operating temperatures have cleared away condensation. Cowl gills, if not fully open prior to starting, must not be adjusted until after a few minutes of engine running.

Provided oil pressure rises satisfactorily, it is desirable to run the engine at the lowest steady speed available for one minute. Then, provided oil inlet temperature has risen sufficiently to prevent damage to the oil cooler and pipe joints, the rpm may be increased to 1,000/1,200 rpm. A safe guide is to run the engine at a rpm that will keep the oil pressure down to 140 lb/sq in. Continue to run at this speed until engine is warmed up sufficiently for ground checks.

GROUND CHECKS (routine): prior to flight the following routine drill must be performed.

i. Set speed control lever at maximum rpm. Adjust throttle to give zero boost. Check rpm is within 50 rpm of that particular engine recorded rpm.
ii. Increase engine rpm to 2,400 rpm and retract the propeller speed control lever enough to cause a drop of rpm. Depending on the propeller installed this will be from 400 to 650 rpm drop. Return lever until 2,400 rpm is regained and leave in this position.
iii. Adjust throttle to set 1,200 to 1,500 rpm. Change to 'S' gear and note momentary drop in oil pressure. Check to ensure proper engagement of the 'S' gear clutches by opening the throttle to RATED and verifying that the correct boost is obtained with no fluctuation. Change back to 'M' gear where a similar drop should occur.
iv. Set speed control lever at maximum rpm position. Open throttle to give maximum economical cruising rpm. Move mixture control to WEAK and note the drop in rpm. Return lever to NORMAL.
v. Check boosts at CRUISING, RATED, and TAKEOFF. Throttle back to RATED and check for decrease in rpm. If no decrease in rpm, throttle back further and switch off

each magneto in turn. Rough running or a drop of more than 50 rpm will indicate faulty ignition.

SHUTTING DOWN: run the engine at 800/1,200 rpm for two minutes. Where twenty-degree counterweight propellers are fitted, set the control fully back.

STOPPING THE ENGINE
If a separate cut-out control is fitted,

   i. Fully close the throttle
  ii. Pull the cut-out control out, hold it until engine stops
 iii. Switch of the ignition
  iv. Release cut-out control smartly to return it to the running position
   v. Turn off the fuel

If the cut-out mechanism is interconnected with the engine master fuel cock,

   i. Fully close the throttle
  ii. Turn off the engine fuel cock
 iii. When engine stops, turn off the ignition

OIL DILUTION
It should be operated if a ground temperature below 0°C (32°F) is likely before next run.

   i. Shut down in usual way
  ii. Top up the oil tank
 iii. When the oil inlet temperature is 40 C to 20 C restart the engine
  iv. Run the engine at 800 rpm for four minutes* pressing oil dilution button
   v. Stop the engine, release oil dilution only when engine totally stopped
  vi. Ensure oil dilution button returns to off position

If the engine has not been flown since the last oil dilution the procedure above should be followed except for (iv), run at 800 rpm for only two minutes.*

Oil dilution cleans the engine internals and filters should immediately be inspected and cleaned.

AFTER STOPPING THE ENGINE
Fit protective covers over the engine intake if the engine is to remain idle for any length of time.

Prior to the first run of an engine that has been idle for some time or newly installed,

i. Examine for fuel or oil leaks
ii. Check external nuts are tight and safely locked
iii. Examine sump oil filter. If any foreign matter is present, flush out the oil system with fuel or flushing oil and refill with clean oil. Run engine again for further check.

IN FLIGHT
Correct operation of the engine is very important to get the best performance and longevity.

OIL PRESSURE: should remain at 80 lb/sq in during flight. Pressure may fall at high temperatures and rpm below 2,000 but if less than 60 lb/sq in above 2,000 rpm the engine should be feathered. If oil inlet temperatures rise rapidly to over 80°C (176°F) an emergency limit of 50lb/sq in is permissible for five minutes. A tolerance of +/- 5lb/sq in is allowed for possible cockpit gauge inaccuracies.

OIL TEMPERATURE: in normal operations should be in 55°C to 65°C range (131°F to 149°F). A maximum oil temperature of 100°C (212°F) is permissible in an emergency for short periods.

---

* One minute for Bristol Beaufighter VI, X and XI for both conditions

CYLINDER TEMPERATURE: the cowl gills must be adjusted to keep the cylinder head temperature within limits. Correct manipulation of the engine controls will keep the temperatures within limits. If a temperature of 300°C (572°F) has been equaled for five minutes, or exceeded, it must be reported.

EXACTOR: the Exactor controls should be primed at frequent intervals. Failure to prime may result in increased fuel consumption owing to incorrect engine conditions brought about by creep of the throttle, mixture, or propeller controls.

AUXILIARY FUEL PUMPS: the electric pumps should be operated as per the instructions for the particular aircraft and fuel system.

CONSTANT-SPEED PROPELLER CONTROL: for taxiing and manoeuvring on the ground, max operating conditions, and takeoff and initial climb, unless maximum power is not required, set the engine speed control in the maximum rpm position. For approach and landing, the control should be set to the maximum climbing rpm. This setting will generally be required for all climbing and maximum cruising. Lower rpm will normally be used for cruising. During a dive the speed controller may be allowed to remain in the cruise position, but high-speed dives with the throttle less than one-third open is forbidden.

    i. Feathering, use the feathering mechanism, immediately close the throttle, and use cut-out. Then turn off the engine fuel cock and switch off the ignition when the engine has stopped rotating.
    ii. Unfeathering, check throttles is closed and engine speed control lever is in the minimum constant-speed rpm position. Keep the cowl gills closed, switch on the ignition, and turn on the appropriate fuel cock. Operate the feather control switch to unfeather depending on the propeller

installed. Always start the engine as slow as possible to prevent damage to the oil pipe connections and engine cooler through excessive oil pressure. Gradually open the throttle to warm up the engine and then adjust the speed control to match the other engines.

TWO-SPEED SUPERCHARGER OPERATION: only applies to engines that the control is not locked in the 'M' gear. The height at which an increase in power can be obtained by changing from 'M' to 'S' gear depends on the boost and rpm. The change should be made as per the following table.

Engine	Condition		
	All-out	Max. Climb and Max. Continuous Cruise	Max. Economical Cruise
XVIII	+5¾ lb	+4½ lb	+¾ lb
VI, XVI (Fully-rated)	+5 lb	+3½ lb	+½ lb
VI, XVI (De-rated)	+2 lb	+½ lb	-1½ lb
XI (Fully-rated)	+4 lb	+1¾ lb	-1¾ lb
XI (De-rated)	+1½ lb	Zero	-1¾ lb

The use of 'S' gear below the altitudes at which the boost has fallen to the values in the table will result in a direct loss of power. For this reason, 'M' gear must always be used for takeoff. Either before, or after landing, the supercharger gear-change mechanism should be exercised to prevent sludge accumulating in the hydraulic clutches. When gear is changed from 'M' to 'S' in flight below 5,000 ft (1,524 m), engine speed should be reduced to 2,000 rpm and the boost to -2 lb/sq in. When exercising after landing rpm should not exceed 1,500 rpm. Oil pressure will drop momentarily when changing gear.

TAKEOFF & CLIMB

i. Before takeoff: to avoid overheating the engine during taxiing the rpm must be kept as low as possible, the cowl gills fully open, and the air-intake heat control set to COLD air. The cylinder temperature should not be allowed to rise above 230°C (446°F) and takeoff is not permitted above this value. If exceeded, the engine must be cooled by running at 8-900 rpm. When possible, avoid taxiing behind other aircraft to prevent ingestion of dust and small stones. If an engine has been idle for more than five minutes, it must be opened up to clear the spark plugs before takeoff. Prior to takeoff set the speed control lever to maximum rpm and the supercharger is in 'M' gear. Set cowl gills in correct position to maintain cylinder head temperature within limits.

ii. Takeoff and initial climb: if back-up pumps fitted in fuel tanks they should be turned on prior to takeoff to prevent possible fuel shortage. In case of lightly loaded aircraft, where maximum power is not required, the throttle should be opened to the TAKEOFF position, and any required reduction in power obtained using an rpm lower than the maximum. This method will ensure that the power and enrichment jets are in operation. Run the engine at maximum conditions for the shortest period consistent with a safe takeoff. As soon as possible after takeoff, set the speed control lever to give the appropriate climbing rpm and retract the throttle to RATED.

iii. Continued climb: during a climb in 'M' gear the boost is maintained by the action of the automatic boost control up to the rated altitude for this gear. At this height the butterfly throttles are fully open so that any further increase in height causes a drop in boost. The supercharger should now be changed to 'S' gear as per the previous 'supercharger gear change schedule' table to get maximum climbing power. When 'S' gear is engaged the climbing boost will be regained and the automatic boost control will maintain it up to the 'S' gear rated altitude. Above this height the boost will start to decrease.

VI & XI engines

These engines have two-lever carburettors and the throttle should be retracted to CRUISING when the boost in 'S' gear has dropped to the maximum economical cruising figure. This will cut the power jets off and prevent loss of power and fuel waste.

XVI, XVII & XVIII engines

These engines have fully automatic carburettors and the mixture strength provided by the larger power jet is no longer required when the altitude has caused the boost in 'S' gear to fall below the maximum climbing figure. If throttle is set at RATED an unnecessary loss of power will result from over-richness, if throttle is retracted to CRUISING the power jet is cut off and economical cruising mixture is automatically obtained. A special power jet is installed to permit operating with an intermediate mixture strength by selecting the throttle halfway between RATED and CRUISING, marked by a white line.

During climb the cowl gills should be adjusted to maintain cylinder head temperatures within limits. It should be possible, under most rated conditions, to climb with cowl gills closed provided the airspeed is adequate. Opening the gills causes drag and waste of fuel. The air intake must be set at COLD AIR during all climbing.

CRUISING

i. Maximum continuous cruise: is the highest power condition that the engine can run for unlimited time. The fuel consumption at maximum cruising power is approximately double the maximum economical cruising figure. The throttle should not be moved beyond RATED setting. If, due to altitude, a boost higher than the maximum permissible for economical cruising is not obtainable, the throttle should not be moved beyond CRUISING. This prevents loss in performance due over-richness. The mixture control of the VI and XI engines should be set to WEAK.

ii. Economical cruising: to save fuel and wear and tear, the 'high boost/low rpm' rule should be strictly observed. The rpm should be reduced consistent with smooth running. To get maximum range set the throttle at CRUISING and maintain optimum indicated airspeed, as per aircraft type, by reducing rpm to the lowest speed consistent with smooth running. For maximum duration, reduce the rpm to minimum and set the boost to the lowest figure at which the aircraft will fly comfortably. Increase speed from this setting by raising boost rather than rpm. A general rule – although the supercharger absorbs more power when operating in 'S' gear than in 'M' gear at the same engine rpm, it is possible under full butterfly-throttle conditions to obtain the same Brake Horse Power at a much lower rpm in 'S' gear than in 'M' gear. The decrease in engine rpm is such that the sum of the friction and supercharger powers is less in 'S' than in 'M', and it is therefore more economical to use 'S' gear in these conditions.

MAXIMUM OPERATING CONDITIONS: the engine must not normally be run at for longer than five minutes at figure higher than those quoted for maximum continuous cruising and never exceed the values in the Operating Data, except rpm in a dive, for the engine. Excessive periods of running at maximum conditions must be reported. The air-intake heat control must be set to COLD Air and cowl gills adjusted as necessary. The throttle must never be closed to CRUISING when the rpm is above the maximum permissible for economic cruising.

DIVING: The engine rpm, except for 20 seconds, and boost in the dive must not exceed the maximum operating figures for a dive in the Operating Data. The speed control lever may be allowed to remain at its previous cruise setting. The throttle should always be at least one-third open to prevent excessive backfiring which could damage the exhaust. Before starting a dive set the supercharger to 'M' gear. The air-intake heat control should be set to COLD AIR and cowl gills fully closed.

LANDING

When landing the cowl gills should be fully closed and the speed control lever set to permit maximum climbing rpm. It may be advantageous at times to place the propeller blades in fine pitch, speed control lever at maximum rpm setting, and windmill the engines for braking effect. Exercise the supercharger and set in 'M' gear. If an engine has been switched off due to imminent failure, do not restart prior to landing in case of sudden failure. After landing open cowl gills. Shut down engines with the aircraft facing in to wind if possible.

MAINTENANCE, DAILY POST-FLIGHT INSPECTION PRECIS

   i. Check the fuel system pipe-lines and unions for leaks
  ii. Check the fuel pump, hydraulic pump, and supercharger volute-casing drain pipes are clear
 iii. Check oil system pipe-lines, unions, rubber connections, and engine in general for leaks
  iv. Check propeller constant-speed governor for security and leaks in the pipe-lines, unions, and the unit itself
   v. Check the ignition high tension leads and spark plug screened connections for security and signs of burning
  vi. Check BTH compressor sump oil level and replenish if necessary, if fitted
 vii. Check the carburettor air-intake shutter control for correct operation
viii. Check the ice guard is not cracked, secure, and not obstructed
  ix. Check cowling gills undamaged, gill plate locking bars flush with support arms, and for correct operation
   x. Check oil cooler for leaks, security, and that the air passages are clear

Bristol Hercules: The Nuts and Bolts

xi. Check inter-cylinder baffles are secure and not chafing
xii. Check oil sump for leaks
xiii. Check oil level in accessory gearbox, replenish as necessary
xiv. Check exhaust pipes and elbows for splits, leaks, and damage
xv. Check nacelles for tools or rags left inside
xvi. Check engine nacelles for damage and that the nacelles and their attachment cables are secure. High oil temperatures have been caused by badly fitting blanking plates allowing hot engine air to circulate through the cooler.

The engines all ran on 100/130 fuel except as noted with DTD 230, 87 Octane fuel. The following table includes the engine number in the series, the Takeoff Brake Horse Power, RPM, and Maximum Boost. A brief description is included.

Engine	Takeoff*	Description
I	1,325; A; +3.25	(1936) Single-speed, medium supercharged 87 octane
II	1,325; A, +3.25	(1938) Single-speed, medium supercharged. Accessories driven by Bristol auxiliary gearbox allowed rapid engine changes 87 oct
III	1,400; A; +4	(1939) Two-speed, full & medium supercharged
III	1,375; A; +4	Two-speed, full & medium supercharged 87 oct
IV	1,380; A; +3	(1939) Single-speed, medium supercharged 87 oct
IVHY	1,380; A; +3	(1939) IV with Hydromatic propeller
V	1,380; A; +3	(1939) Civil version of IV; prototype only; not developed
VI	1,615; B; +8.25	(1941) Two-speed, full & medium supercharged
VI	1,350; A; +5	Two-speed, full & medium supercharged 87 oct

The Hercules: The Other Engine That Helped Win the War

Engine	Takeoff*	Description
VII	1,700; B; +8.25	Two-speed, full & medium supercharged cancelled
VIII	1,360; A; +4	Very high altitude version of II. Single-speed medium supercharger plus auxiliary high altitude Single-speed 'S' supercharger 87 oct
X	1,420; A; +4	(1941) As III with 10 degree supercharger diffuser 87 oct
XI	1,590; B; +5	(1941) As III
XII	1,615; B; +8.25	(1941) Two-speed; full & medium supercharged; As VI
XIV	1,500; A; +3.5	(1942) Civil version of IV for BOAC
XVMT	1,650; A: n/avail.	Very high altitude version of II. Single-speed medium supercharger plus auxiliary high altitude Single-speed 'S' supercharger
XVI	1,350; A; +5	As VI with Hobson single lever carburettor 87 oct
XVI	1,615; B; +8.25	As VI with Hobson single lever carburettor
XVII	1,725; B; +8.25	(1943) As VII with supercharger locked in 'M' gear; reduced 12-inch impeller; single lever carburettor
XVII	1,690; A; +8.25	(1943) As VII with supercharger locked in 'M' gear; reduced 12-inch impeller; and single lever carburettor 87 oct
XVIII	1,700; B; +8.25	As VI, low level cropped 12-inch supercharger impeller
XIX	1,725; B; +8.25	(1943) As XVII with 'Long Tom' plugs & ignition harness
XX	1,700; B; +8.25	As XIX
36	Not available	As VI, XVI with Hobson-RAE injector
38	Not available	As 36 locked in 'M' gear with turbo-supercharger

* Takeoff; Sea Level; Brake Horse Power; RPM A=2,800; B=2,900; Maximum Boost

The following is a numerical list describing the superchargers for the engine series. The 100 was the basis of a new family of engines with civil applications. It had exceptional reliability and economy with longer time between overhauls. The new nomenclature would have 500 added to the military engines to give the civil variant. The series would be split as the series progressed into the Bristol epicyclic reduction gearing, indicated by even numbers, and a new concept, Torquemeter-type reduction gear, by odd numbers.

Engine	Takeoff*	Description
100	1,675; A; +8.25	(HE-10SM) Turbine entry Two-speed supercharger
101	1,640; A; +8.25	As 100 with rear-swept exhaust
103	1,640; A; +8.25	Torquemeter version of 101
105	1,675; A; +7.25	As 100 with 120 supercharger gears
106	1,675; A; +10	As 101 with modified injector and increased boost
107	1,675; A; +10	Torquemeter version of 106
110	1,640; A; +8.25	As 101 with 150 hp accessory drive; submerged scavenge pump; vertically mounted starter
120	1,640; A; +8.25	As 110, high altitude variant with cabin supercharger
121	1,715; A; +8.5	Torquemeter version of 121
130	1,715; A; +8.5	As 100 with four point mounting ring
134	1,690; A; +8.5	As 130 with modified four point mounting ring and rear swept exhaust to suit a rear manifold
200	1,640; A; +8.25	As 120. Basic engine for 200 series. A high altitude variant with cabin supercharger; 150 hp accessory drive; submerged scavenge pump; vertically mounted starter.
216	1,800; A; +12.5	As 106 with 230 power section and single-speed supercharger

The Hercules: The Other Engine That Helped Win the War

Engine	Takeoff*	Description
230	1,925; A; +11.5	As 130 with 1 inch bearing rollers increasing performance
231	1,925; A; +11.5	Torquemeter version of 230
232	1,925; A; +11.5	As 230 with six point mounting ring; small diameter ignition harness; 3 inch exhaust pipes; master connecting rods in cylinders 6 and 7.
233	1,925; A; +11.5	Torquemeter version of 232
234	1,980; A; +13	As 232 with large diameter ignition harness; modified rich mixture rating; free exit cowling; master connecting rods in cylinders 6 and 7.
235	1,980; A; +13	Torquemeter version of 234
238	1,980; A; +13	As 234 (734), a military engine
260	1,925; A; +11.5	As 230 but with 1 inch bearing rollers; redesigned two-speed epicyclic supercharger drive; six point mounting ring; front cover and reduction gear to suit braking-type propeller; submerged oil pipes; 150 hp accessory drive; cylinder heads sealed to suit free exit cowling; rear exhaust manifold
261	1,950; A; +13	Torquemeter version of 260
264	1,950; A; +13	As 260 with 230 front cover and reduction gear. Cylinder heads sealed to suit 'clover leaf' cowling and rear manifold
265	1,950; A; +13	Torquemeter version of 264
268	1,980; A; +13	As 264 with head seals to suit free exit cowling; rear-swept exhaust pipes; increased climb rpm
269	1,980; A; +13	Torquemeter version of 268
270	1,980; A; +11.5	As 230 with a redesigned rear cover having increased power accessory gearbox drive and provision for tachometer generator drive. Modified starter position

Bristol Hercules: The Nuts and Bolts

Engine	Takeoff*	Description
271	1,980; A; +11.5	Torquemeter version of 270
630	Not available	As per 100; 230 with a four point mounting ring; front exhaust system; single-speed; medium supercharged civil- series engine
631	1,980; A; +11.5	Torquemeter version of 630
632	1,600; A; +8.5	As 630 with six point mounting ring and rear swept exhaust system. A civil series engine
633	1,600; A; +8.5	Torquemeter version of 632
634	1,600; A; +8.5	As 630 with modified four point mounting ring and exhaust pipes swept to rear manifold. A civil series engine
635	1,600; A; +8.5	Torquemeter version of 634
636	Not available	As 630 with special mounting attachments for installation in a monocoque engine nacelle; modified magnetos; master connecting rods in cylinders 6 and 7
637	1,690; A; +8	Torquemeter version of 636 with front exhaust system
637-2	1,690; A; +8	Torquemeter version of 636 with increased oil consumption limits and different piston rings assembly
637-3	1,690; A; +8	Torquemeter version of 636 with copper-based cylinder heads; modified cylinder head baffles; a larger oil pump
638	1,690; A; +8.5	As 632 with rear swept exhaust pipes. A civil series engine
639	1,690; A; +8.5	Torquemeter version of 638
656	Not available	As 636 with front cover and reduction gear to suit braking-type propeller
657	Not available	Torquemeter version of 656 with front exhaust system

The Hercules: The Other Engine That Helped Win the War

Engine	Takeoff*	Description
672	1,690; A; +8	As 632 with fairings for free exit cowling and rear-swept exhaust pipes. A civil series engine
673	1,690; A; +8	Torquemeter version of 672
		ALL 700 SERIES ENGINES ARE CIVIL VARIANTS
730	2,040; A; +13	As 230/630 with redesigned power section; 1 inch bearing rollers; six point mounting ring; free exit cowling; rear manifold; small diameter ignition harness; modified magnetos; master connecting rods in cylinders 4 and 11
731	2,040; A; +13	Torquemeter version of 730
732	2,040; A; +13	As 730 with master connecting rods in 6 and 7
733	2,040; A; +13	Torquemeter version of 732
734	2,040; A; +13	As 730 with large diameter ignition harness and standard magnetos
735	2,040; A; +13	Torquemeter version of 734
736	2,040; A; +13	As 730 with modified six point mounting ring; parallel cowling; large diameter ignition harness; standard magnetos
737	2,040; A; +13	Torquemeter version of 736
738	2,040; A; +13	As 730 with rear swept pipes; large diameter ignition harness; standard magnetos
739	2,040; A; +13	Torquemeter version of 738
750	Not available	As 730 with redesigned front cover and reduction gear to suit braking-type propeller Rear exhaust pipes
751	Not available	Torquemeter version of 750
758	2,040; A; +13	As 750 with free exit cowling and rear swept pipes
759	2,040; A; +13	Torquemeter version of 758

Bristol Hercules: The Nuts and Bolts

Engine	Takeoff*	Description
760	Not available	As 730 with two-speed epicyclic drive supercharger; front cover and reduction gear for braking propeller; submerged oil pumps; 150 hp accessory drive; rear manifold exhaust pipes
762	2,040; B; +16.5	As 760 with modified supercharger to give high altitude rating using 115/145 fuel
763	2,040; B; +16.5	Torquemeter version of 762
772	1,965; A; +13.75	As 762 with two-speed supercharger locked in 'M' gear. Supercharger impeller cropped to suit 100/130 fuel with methanol/water injection
773	1,965; A; +13.75	Torquemeter version of 772
790	2,040; A; +13	As 758 with redesigned rear cover; modified starter; a 150 hp accessory gearbox drive; provision for tachometer generator
791	2,040; A; +13	Torquemeter version of 790

* Sea Level; Brake Horse Power; RPM A=2,800; B=2,900; Maximum Boost

The following is a list of superchargers for the engine series.

Engine	Code	Supercharger
I	HE-IS50	fully supercharged.
I	HE-IM50	medium-speed supercharged.
II	HE-IM	medium supercharged
III	HE-ISM	fully/medium supercharged
IV	HE-4M	single-speed; medium supercharged
VI	HE-6SM	two speed, fully medium supercharged
VII	HE-7SM	two-speed, fully/medium supercharged
VIII	HE-8MAS	single-speed; medium supercharged with auxiliary high-altitude; single-speed 'S' blower

The Hercules: The Other Engine That Helped Win the War

Engine	Code	Supercharger
X	HE-1SMB	fully/medium supercharged with 10 degree diffuser.
XVII	HE-7SM	two-speed, fully/medium supercharged
36	HE-9SM	two-speed, fully/medium supercharged
38	HE-11MT	medium supercharged, medium turbo.
100	HE-10SM	turbine entry two-speed supercharger
101	HE-10SM	turbine entry two-speed supercharger
110	HE-18SM	turbine entry two-speed supercharger
120	HE-18SM	turbine entry two-speed supercharger
130	HE-10M	turbine entry two-speed supercharger
200	HE-18SM	turbine entry two-speed supercharger
230	HE-20SM	turbine entry two-speed supercharger

The Hercules was not only installed in military aircraft, it was suitable for civil designs as well. The Hercules engines continue to fly in the twenty-first century in a private society Nord 2501 Noratlas based in France. The engine and airframe combinations listed below are deemed, as far as research permitted, to have been recorded as flown together. This would include 'one-off' engine testing and experimental trial installations that never went into production. The basis of the list is Alec Lumsden's *British Piston Aero-Engines and Their Aircraft*. I am indebted to his work.

Aeroplane	Series
Armstrong Whitworth A.W.41 Albermarle I	III
Armstrong Whitworth A.W.41 Albermarle I; II; V; VI	X; XI
Avro 683 Lancaster II	VI; XVI
Avro 685 York II	XVI
Avro 689 Tudor VII	120
Blackburn & G.A.60 'Universal' Freighter I*	260
Blackburn & G.A.60 'Universal' Freighter II*	730
Breguet 890H prototype	738

Bristol Hercules: The Nuts and Bolts

Aeroplane	Series
Bristol 156 Beaufighter I	III; X; XI
Bristol 156 Beaufighter I; VIC; VIF; XTF	VI
Bristol 156 Beaufighter I; VIC; VIF; TFX	XVI
Bristol 156 Beaufighter TFX; XIC	XVII
Bristol 156 Beaufighter TFX; XIC; 21	XVIII
Bristol 156 Beaufighter VIC	130
Bristol 156 Beaufighter I F.17/39	II
Bristol 170 Freighter	234; 238; 638
Bristol 170 Freighter (R.I.A.F.)	268
Bristol 170 Freighter I; IA; II; IIA; IIB; IIC; XI; XIA	632
Bristol 170 Freighter 21; 21E; 21P	672
Bristol 170 Freighter 31 srs; 32	734
Bristol 170 Wayfarer	632; 638
C.A.S.A. 207 Prototype	730
Fairey Battle I**	II; XI
Fokker T.9	II; XI
Folland Fo.108 43/37	VIII; XVMT
Handley Page H.P.61 Halifax III	VI; XVI
Handley Page H.P.61 Halifax VI	XVI; 100
Handley Page H.P.67 Hastings C.1	100; 101; 105
Handley Page H.P.67 Hastings C.2	106; 216
Handley Page H.P.94 Hastings C.4	736
Handley Page H.P.95 Hastings C.5	736
Handley Page H.P.68 Hermes I	100; 101
Handley Page H.P.74 Hermes II	762
Handley Page H.P.81 Hermes IV	762
Handley Page H.P.81A Hermes IVA	772
Handley Page H.P.69 Halifax VII	VI; XVI
Handley Page H.P.70 Halifax VIII	100
Handley Page H.P.70 Halton I; II***	100
Handley Page H.P.71 Halifax IX	XVI
Handley Page H.P.74 Hermes II	106; 120; 130

The Hercules: The Other Engine That Helped Win the War

Aeroplane	Series
Northrop 2-L Gamma Commercial	I
Nord 1400	100; 739
Nord 2501 prototype	739
Nord 2501	758
Nord Noratlas	790
Saro A.36 Lerwick I	II; IV
Short S.26 'G' class flying-boat	IV; IVHY; XIV
Short S.29 Stirling protype	I
Short S.29 Stirling I	X
Short S.29 Stirling I srs I	II; III
Short S.29 Stirling I srs II	XI
Short S.29 Stirling III; IV; V	VI; XI; XVI
Short S.45 Solent I	XX; 232
Short S.45 Solent 2	636
Short S.45 Solent 3	637-2; 637-3; 656
Short S.45 Solent 4	732
Short S.45 Sunderland IV (Seaford) R.8/4	XIX
Sud Est S.E.1010	101
Vickers 401; 491; 495; 496 Viking I	130
Vickers 637 Valetta C.1	230; 270
Vickers 668 Varsity T.1	264
Vickers 498 Viking I	630
Vickers 498 Viking IA	134
Vickers 610; 614; 616; 621; 623 Viking IA; B	634
Vickers 289 Wellesley	I
Vickers 299; 417 Wellington III	III; X; XI
Vickers 407; 426 Wellington V	VIII; XVMT
Vickers 428 Wellington D.W.I Mk III	XI
Vickers 437 Wellington X	XVI
Vickers 439;440 Wellington IX; X	XVI
Vickers 440 Wellington X	XI

Bristol Hercules: The Nuts and Bolts

Aeroplane	Series
Vickers 454; 458 Wellington XI	VI
Vickers 455 Wellington XII	VI
Vickers 478 Wellington X	100
Vickers 478; 487 etc Wellington X; XI; XII; XIII; XIV; XVII; XVIII	XVI
Vickers 466 etc Wellington X; XI; XII; XIII; XIV; XVII; XVIII	XVII

* Later became Blackburn Beverley, first flown 20 June 1950 with Centaurus engines
** Engine test bed
*** Postwar Halifax bomber conversion to civilian transport

In Chapter 6, we will see what type of aircraft the Hercules was installed in. The Bristol Hercules and Centaurus were at the end of the piston engine era, only to be superseded by the jet era. In fact, both engines overlapped the time of transition as the jet engine rose to prominence.

5

THE HERCULES AND AIRCRAFT COMPANIES

Shortly after its first flight in 1936, the Bristol Hercules engine was installed by the Bristol Aeroplane Company in a Northrop Model 8A-1 test bed aeroplane. The Hercules would thereafter be installed by fifteen different aviation manufacturers, American, European, and British, for test purposes, trial research, and production. The engine would be in service from 1939 until May 2022, the last known flight (prior to manuscript submission) of the private, Association le Noratlas de Provence, 1956 Nord Noratlas 2501F in France. Hopefully, it will continue to be part of the airshow circuit for years to come. The engine design is eighty-three years old and going strong, an inspiration for us all!

These are the companies that used the over fifty-seven thousand Bristol Hercules engines as their powerplant, some only for test bed purposes, some for exclusive type production, and some used in conjunction with other manufacturer's engines in a certain model of aircraft. The horsepower of the initial Hercules engine had increased by thirty-five per cent by the end of the war years. It is understandable that British companies would use the engine during the war, but it is somewhat surprising that foreign companies, such as Nord, would use it in the 1950s; well, upon reflection, maybe not. The even more powerful Hercules in that era was still proving its worth in post-war transport aircraft, such as those of Handley Page, and the French aviation industry was still rebuilding after the ravages of war on its soil. Depredations were not only in the physical structures, factories and engine test facilities, but in the loss of qualified aviation engine personnel.

Armstrong-Whitworth

Sir W. G. Armstrong Whitworth & Co Ltd
Elswick, Newcastle-upon-Tyne
Founded: 1847, by William Armstrong

Armstrong founded the Elswick works in 1847 to produce cranes, bridges, and hydraulic machinery. He progressed into artillery with a breech loading gun that saw service in the Crimea War. The company merged with the shipbuilding firm of George Mitchell to form Armstrong Mitchell & Company on the banks of the River Tyne. This in turn merged fifteen years later with the engineering firm of Joseph Whitworth and at the beginning of the twentieth century manufactured cars and trucks.

The Aerial Department was formed in 1912 as part of Sir W. G. Armstrong & Company engineering Group in Newcastle-upon-Tyne. During the first few years of the First World War, it employed the services of the Dutch aircraft designer Frederick Koolhoven, hence the F.K. designation for the early aircraft models. The engine and automotive parts of the company separated and the aviation portion remained as the Sir W. G. Armstrong Whitworth Aircraft Company. Out of the eleven wartime aircraft built, the general purpose biplane, FK 8, was the most widely produced. In 1920, this part of the company again changed its name and became the Armstrong Whitworth Aviation subsidiary. A merger with Vickers Limited followed in 1927 to form Vickers-Armstrongs. This left two companies, Vickers-Armstrong, commonly referred to as Vickers, and Armstrong-Whitworth, which produced nearly thirty aircraft

The Hercules: The Other Engine That Helped Win the War

before the AW 38 Whitley in 1936 and the AW 41 Albermarle in 1940 were both ordered in substantial numbers for the RAF. The company produced over 1,300 Lancaster bombers during the war. In 1959, the AW 650/660 was a successful transport/cargo aircraft and the last design to be produced. In 1961 the company became Whitworth Gloster Aircraft and in 1963 the name disappeared in another merger, with Hawker Siddeley.

As for the automobile component, the end of the Boer War caused a shift in company focus to the emerging car market. It took over the Wilson-Pilcher design, a 9 hp flat, four-cylinder engined automobile, and continued to develop the design until 1906 when it produced its first Armstrong Whitworth model. Further automotive development brought reliability and solid workmanship features. By 1919 it was an attractive company and was taken over by Siddeley-Deasy to become Armstrong-Siddeley.

The major shipbuilding division of the company started life as Charles Mitchell shipbuilding, to become Armstrong Mitchell, which produced many navy and merchant ships to meet a variety of needs for a wide range of customers. It constructed a cruiser for the Chilean Navy and followed up with warships for the Royal Navy, Imperial Russian Navy, Imperial Japanese Navy, Brazilian Navy, Spanish Navy, and the United States Navy. Gunboats were produced for the Italian Navy and South Australian Colonial Navy. The merchant ships included oil tankers, freighters, dredgers, tankships, and speciality vessels such as the first polar ice breaker with a strengthened hull to ride over and break up the ice.

The Elswick Ordnance Company was a major supplier of ordnance and ammunition during the First World War to its main customer, the British Government. The hydraulic engineering arm produced a series of cranes and bridges, the most famous being Tower Bridge, London in 1894. The company supplied locomotives to Britain and many markets overseas, both steam and diesel engines after the First World War. The overseas railway customers included Bombay, Baroda and Central India, Madras and Southern Mahratta, Buenos Aires Western, Belgian State, East Indian, South Australian, Bengal Nagpur, Queensland, Egyptian State, and Ceylon Government railways. Two hydroelectric stations were built abroad, one completed in 1925 at Deer Lake, Newfoundland and the other in 1924 at Nymboida, New South Wales, Australia.

Avro

A. V. Roe and Company
Brownsfield Mill, Great Ancoats Street, Manchester
Founded 1910, became Hawker Siddeley Aircraft in 1963

Notwithstanding the fact that Alliott Verdun Roe and his brother Humphrey founded A. V. Roe and Company on 1 January 1910, Avro should be synonymous with Roy Chadwick. Together, they made a strong team which contributed so much to the British aircraft industry with their successful designs. Humphrey was the business brains as the Managing Director until he joined the Royal Flying Corps in 1917 and subsequently went his own way in the medical field. Alliott had already set an important milestone by designing and flying the Roe I triplane, the first all-British powered flight. Two years later, Roy Chadwick joined the company as draughtsman and personal assistant to Alliott and in 1918, based on his performance, he became the Chief Designer. His innovative ideas started with the alphabetic series, A, D, F, and the 500 numbered series, 500, 501, and 503. This culminated in the very successful Avro 504 light bomber and trainer with over 8,000 aircraft being built in Britain.

Post-war, the company ran into financial problems and a majority share was bought by Crossley Motors to enable it to have factory space for automotive body building. However, Avro still produced aircraft such as the 1926 Avro Avian, which two years later was flown solo from England to Australia. In that year, Avro was sold to Armstrong Siddeley

The Hercules: The Other Engine That Helped Win the War

Holdings and Alliott resigned from the company and founded the Saunders-Roe company. In 1935, Chadwick designed the very successful multi-role and trainer aircraft, the Avro Anson. Over 11,000 Ansons were built. The same year, Avro became a subsidiary of Hawker Siddeley. The political situation in Europe changed the focus of the company from trainer aircraft to bomber aircraft.

The Avro Manchester became the foundation of the most famous of all the Second World War bombers, the Avro Lancaster. The Lancaster BII had Bristol Hercules engines. During this same era Chadwick designed the Lincoln bomber and York transport. The 'shadow' factory close to Leeds Bradford airport was the largest in Europe, employing more than 17,000 workers in shifts.

In 1945, A. V. Roe Canada Limited was founded at the site of the Victory Aircraft factory near Toronto, Ontario. It, too, was a subsidiary of the Hawker Siddeley Group and used the name Avro Canada. The author flew one of their products in the 1970s, the Avro CF-100 Mk 5C/D, for the Canadian Armed Forces, now known as the Royal Canadian Air Force. Postwar, Avro developed the civil Lancastrian transport and the maritime reconnaissance Shackleton aircraft. Roy Chadwick was tragically killed in the Avro Tudor pressurised airliner prototype in August 1947.

Avro transitioned into the jet age with the nuclear-strike Vulcan bomber armed with its own Blue Steel standoff nuclear bomb. The Vulcan would be followed by the Avro 748 twin turboprop airliner. By now, Avro had been absorbed into Hawker Siddeley Aviation and the name ceased to be used. The name Avro would later be resurrected in a series of regional jets; such was the strong heritage feeling around the name. Perhaps some idea of the scope of Avro can be gained by the sequential numbering of aircraft proposals/projects, some successful, some not, and some unbuilt. The series ran from 500 to 784. The list included such aircraft as the Avro 504, Avro 652 Anson, Avro 683 Lancaster, Avro 694 Lincoln, and Avro 698 Vulcan.

Blackburn and General Aircraft Limited

Brough, Yorkshire
Founded: 1914 became Hawker Siddeley in 1960

Robert Blackburn started the Blackburn Aeroplane and Motor Company Limited in premises at the Olympia Roller-Skating Rink, Roundhay Road, Leeds. Two years later a new factory was built at Brough, East Riding of Yorkshire. By 1933 all production was at Brough. In 1929, an American branch was incorporated, Blackburn Aircraft Corporation in Detroit, Michigan. In 1934, Blackburn acquired the Cirrus-Hermes Engineering company but retained it as a separate company until it had proven itself. In 1936, the company changed its name to Blackburn Aircraft Limited. Finally in 1937, the engine manufacturing business was brought in to the parent company under the Blackburn Cirrus name.

Once again, the pressures of a looming war created a fortuitous business climate and a new factory was built at Barge Park, Dumbarton, Scotland. This ideal location was adjacent to shipbuilding facilities with launch areas. The Blackburn Shark, a biplane torpedo bomber/reconnaissance aircraft, was produced and underwent modification at the William Denny and Brothers' facility in Dumbarton. A total of 241 Short Sunderland four-engine flying boats that were used for long-distance transport, reconnaissance, U-boat patrols, and search and rescue were produced at Dumbarton from October 1941 until October 1945. At its peak the factory employed 4,000 workers. Blackburn also produced other aircraft manufacturers' aircraft, such as the Fairey Swordfish, at their Sherburn-in-Elmet factory, East Yorkshire.

In 1949 Blackburn amalgamated with General Aircraft Limited to become Blackburn and General Aircraft Limited, reverting in 1958 to Blackburn Aircraft Limited. Post-war, Blackburn continued its engine component operations and combined with Turbomeca to produce gas turbine engines.

The aircraft component of the company was absorbed into Hawker Siddeley in 1960 and the name was no longer used after 1963. There were over eighty proposed projects. Some were produced for the First World War, an example being the Blackburn Kangaroo, and some for the Second World War, such as the Blackburn Botha, followed by the Blackburn Beverley transport, with Bristol Hercules engines, and Blackburn Buccaneer naval strike aircraft in the 1950s.

Bristol

Bristol Aeroplane Company
Filton, Bristol

Founded in 1910, the company split in 1956 into Bristol Aircraft and Bristol Aero Engines. In 195,9 Bristol Aircraft became the British Aircraft Corporation and Bristol Aero Engines merged with Armstrong Siddeley to form Bristol Siddeley. The Bristol Aeroplane Company is described in Chapter One.

Breguet

Breguet Aviation
La Brayelle, Douai, France
Founded 1911, became Dassault Aviation in 1971

La Société anonyme des ateliers d'aviation Louis Breguet, commonly referred to as Breguet Aviation, was founded by the aircraft designer Louis-Charles Breguet and his brother Jacques. In its first year of production their aircraft established a speed record over a ten-kilometre (6.2-mile) course, an auspicious start. This was followed in 1912 by a seaplane project. During the First World War the company produced a range of aircraft including a reconnaissance type which continued in service into the 1920s. Nearly 8,000 Breguet 14s were built to serve as a biplane bomber and reconnaissance aircraft. It was unique in that it had large amounts of aluminium in its construction. It was employed by more than seventy escadrilles and also served in numbers with the forces of the United States Army Air Force, Belgian Army, China, Brazil, Finland, Greece, and Spain. The Michelin industrial conglomerate also helped to produce the Breguet 14 for the war effort and beyond.

During the 1940s Breguet was involved in the innovative design and production of an electric car. It succumbed to the exigencies of war. Post-war, Breguet was very active, with long-distance record-setting flights and pioneering work on such ideas as a gyroplane. It turned its attention to transport type aircraft to meet the new aviation requirements. In 1949, the prototype Mercure had the Bristol Hercules engines. The Breguet 763 was a double-decker transport that had four Pratt & Whitney engines. It was flown by Air France and the French Air Force. In response to a North Atlantic Treaty Organisation specification, Breguet created the Br1150 long-range, maritime patrol aircraft. Renamed the 1150 Atlantic, a consortium was set up to produce it from 1965 to 1974. It used the Rolls-Royce Tyne engine and was a successful project that was updated as the Atlantique 2, starting in 1989 with advanced avionics and equipment.

Breguet was a part of SEPECAT, a multi-national joint venture company, in the 1960s, that built the Jaguar strike aircraft in association with the British Aircraft Corporation. The aircraft was fully assembled in France and Great Britain. Shortly after, in 1971, Breguet was merged with Dassault Aviation, headquartered in Paris, and emphasis shifted to the in-house products of the Super Etendard naval attack aircraft, and Mirage F1 interceptor aircraft.

CASA

Getafe, Cadiz, Alicante, Sabadell, Tablada, Spain
Founded 1923, became EADS CASA in 1999 and Airbus Military in 2009

The Hercules: The Other Engine That Helped Win the War

The Spanish company CASA had a connection with Breguet. Construcciones Aeronauticas SA, began work building Breguet aircraft in Getafe under licence. Subsequently, in 1926, it built the German Dornier DO. J Wal seaplane for the Spanish Air Force. This was followed by the first of its own CASA designs in 1929. The CASA-1 was the prototype for the CASA III to follow. In 1932, the company obtained a licence from the British aircraft company Vickers to build the Vickers Vildebeest land-based torpedo bomber. A truly international company, CASA built the Russian Polikarpov I-15 biplane fighter. The Spanish Civil War disrupted production and factories were moved to avoid the hostilities. A new factory in Tablada built various German aircraft under licence, such as the Gotha Go 145A (CASA 1145), the Bucker Bu 133 (CASA 1133), and over 500 Bucker Bu 131 (CASA 1131) aircraft.

By 1945, CASA was manufacturing a two-engine bomber, the Heinkel He 111 CASA 2.111, but with Rolls-Royce Merlin engines installed. The Spanish Government increased their investment in the company, which allowed other components to grow, such as the Design Office. By 1957 CASA had a contract for the NATO F-100 Super Sabres based in Europe and the Spanish Air Force North American T-33. These contracts were followed by the production of the CASA 207 B with Bristol Hercules engines in the 1960s and the Northrop F-5A fighter-bomber in 1962.

In 1971 CASA merged with Hispano Aviacion. The following year it became one of the founding members of the Airbus Consortium and in 1996 joined the Eurofighter 2000 project. CASA has one of the assembly lines for the Eurofighter Typhoon. Ever expanding, it joined the European aerospace corporation (EADS) in 1999. EADS-CASA had 7,500 employees as part of the 100,000 strong EADS conglomerate of Aerospatiale-Matra of France and Daimler Chrysler Aerospace of Germany. In 2009, the name finally disappeared, it is now part of Airbus Military.

The Hercules and Aircraft Companies

Fairey Aviation Company

Hayes, Middlesex; Heaton Chapel & RAF Ringway, Cheshire
Founded 1915, became Westland Aircraft in 1960, now WFEL Limited, Spectris

One company's loss can be another company's gain. Such is the case with Richard Fairey and Ernest Tips, former employees of Short Brothers, who decided to form a manufacturing company, the Fairey Aviation Company. Initially, the company built other manufacturer's aircraft before turning their attention to naval aviation. By 1917, they had designed an aircraft specifically for use as a patrol and reconnaissance aircraft flown from an aircraft carrier. It was a single-engine, two-seat biplane with two main floats and backwards-folding wings. The Fairey Campania was the first aeroplane to be designed solely for carrier operation; the Royal Navy Air Service was leading the way. Over sixty Campanias were built, initial models powered by the Rolls-Royce Eagle engine. The main production facility was at Hayes, Middlesex, with flight testing at Northolt Aerodrome until 1930 and then Great West Aerodrome until shortly before the end of the Second World War. Great West became London Heathrow Airport after expropriation by the Air Ministry in 1944. Subsequent testing was at Heston and White Waltham airports. Co-located at Hayes was the propeller division.

In the mid-1930s the facilities were expanded to Stockport, Manchester, and flight testing was done in 1937 at Ringway (Manchester) Airport. Over forty Fairey aircraft were designed, including the Swordfish of the early 1930s. The Fairey Battle was built and flown at these locations and was one of the aircraft used as the Hercules test bed. The Battle was followed by the Fulmar fighter and Barracuda dive-bomber during the Second World War. Fairey also built the Bristol Beaufighter and Handley Page Halifax as part of the war effort. Post-war the carrier-borne Firefly and Gannet were used by the Fleet Air Arm.

Fairey Aviation reorganised during the 1950s into aviation and engineering components. In 1960 Westland acquired the aviation component. In 1977, the Fairey Engineering group went into receivership. It should be mentioned that in the 1930s there was a Belgian subsidiary and in 1948 Canadian and Australian subsidiaries were formed. In 1954 the Fairey Delta 2 research aircraft reached a record breaking 1,822 kmh (1,132 mph) in level flight.

Folland

Hamble, Hampshire England
Founded 1936 as British Marine Aircraft Limited, Hawker Siddeley 1959, defunct 1963

The company was formed specifically to produce under licence the long-range, four-engine Sikorsky S42A flying boat. A factory was built with access to Southampton Water on the western side of the Hamble peninsula. Production started but was short-lived as British Marine ran out of funds and went into liquidation. Fortuitously, Henry Folland, Chief Designer at Gloster Aircraft, had resigned and was looking for facilities to set up his own production company. In 1937 he started making parts for the Bristol Blenheim and Beaufort bombers. The demands of a looming military confrontation provided further work for Folland with the Supermarine Spitfire, De Havilland Mosquito, and Vickers Wellington. It was not until 1940 that the first Folland-designed aircraft appeared, the Folland Fo 108. This was an aircraft specifically designed to be an engine test bed. It was a large 7,257 kg (16,000 lb), low-wing, cantilever monoplane with a unique cockpit area in that the pilot was enclosed in a glazed area with a cabin behind and below for two observers. This cabin was outfitted with instrumentation so that the performance of the engine could be observed and recorded in flight. The Bristol Hercules engine was used to ferry the aircraft to the new location, such as the Napier, Bristol, and Rolls-Royce engine facilities, for testing of their engines. It was not surprising that five out of the twelve aircraft built were lost to accidents testing new engines.

In 1950, the former English Electric designer, W. Petter, joined the company and embarked on a proof-of-concept project to design a lightweight, swept-wing fighter. The aircraft was destroyed in an accident, but the concept grew into the successful Folland Gnat. It was May 1965 before the Gnat was delivered to the RAF as an advanced trainer. It achieved a highly visible profile by being selected by the RAF's Red Arrows' display formation team. The fighter version was exported to Finland, India, and Yugoslavia as a fighter. The Gnat was flight tested at the former RAF Chilbolton. The premises continued to be used by British Aerospace for components of the Harrier and Hawker jets.

Fokker

Schwerin, Germany; Amsterdam, Holland
Founded 1912, bankrupt 1996

Fokker was a Dutch aircraft manufacturer named after its founder, Anthony Fokker. Fokker moved to Germany in 1912 for better aviation opportunities, such as selling his first aircraft, the Fokker Spin, to the German government. It initially had several name changes, Fokker Aeroplanbau, and Fokker Aviatik GmbH. A factory was set up to supply the German Army for the First World War. The Fokker M5 quickly became the Fokker Eindecker with the addition of a synchronised gun. A group of Fokker fighters became well known among the Allied forces: the Fokker D.VI, D.RI Dreidecker (Red Baron), and D.VIIs. Post-war, Fokker returned to the Netherlands with ample aircraft parts to resume business. The 1920s and 1930s were definitely the glory days for the company. The F.VIIa trimotor passenger aircraft captured forty per cent of the American market before the Ford trimotor appeared. The Bristol Hercules was used in a prototype Fokker T.IX twin-

The Hercules: The Other Engine That Helped Win the War

engine bomber in 1939, but further development stopped with the German invasion of Holland.

Post-war, transport conversions and a trainer for the Royal Netherlands Air Force occupied the company until a transition in 1951. New facilities were built at Schiphol Airport to build the Gloster Meteor and Lockheed Starfighter under licence. In 1958, its first turboprop airliner was produced, the Fokker F27 Friendship. Powered by two Rolls-Royce Dart engines, it became the world's bestselling turboprop airliner selling 800 aircraft by 1986. It was also produced in the US by Fairchild under licence. In 1962, the Friendship was followed by the rear twinjet F28. This was followed by the F16 Fighting Falcon European consortium. Subsequent developments of the F28 led to the F50, F70, and F100. These aircraft were not successful enough to compete against the Boeing and Airbus products on the market. In March 1996, the company was declared bankrupt. The contribution of Fokker to aviation should not be underestimated. Between 1912 and 1996 Fokker created 160 designs, some successful, some not.

Handley Page

Handley Page Aviation Company
Radlett Aerodrome, Hertfordshire
Founded 17 June 1909, voluntary liquidation in 1970

Founded by Frederick Handley Page, it was the United Kingdom's first publicly traded aircraft manufacturing company. Page started his career as an electrical engineer before joining the Royal Aeronautical Society in 1909. He set up his own company and experimented with gliders and his first, unsuccessful, aircraft, named the Bluebird. This pioneering company established their premises at Cricklewood, London, and their aircraft flew from the company's adjacent airfield, Cricklewood Aerodrome. The Handley Page company established a niche for building large,

The Hercules and Aircraft Companies

heavy bombers during the First World War. The O/100 was quickly followed by the O/400 and O/1500 bombers. The company did some experimentation with leading edge wing slots to improve lift at high angles of attack and so delay a stall. These bombers were designed for the Royal Navy to bomb the German Zeppelin yards. The company's airline, Handley Page Transport Ltd, used converted Handley Page bombers for their passenger service in 1920 on a London to Paris route. In 1924, Transport's assets were merged with three other airlines to found Imperial Airways, the United Kingdom's national airline service. Handley Page Transport provided a world first: an in-flight meal. It continued to develop large transport aircraft and its Handley Page HP 42 aircraft was used on overseas routes.

In 1929, Cricklewood was closed and flying operations moved to Radlett Aerodrome in Hertfordshire. With war looming, a Handley Page aircraft joined two other medium bombers, the Armstrong Whitworth Whitley and Vickers Wellington, the HP 52 Hampden, which was first flown in 1936. The government released a specification for a heavier twin-engine bomber and Handley Page responded with the Rolls-Royce Vulture engines. The Vulture engines had too many teething problems and the effort was abandoned. The specification became the four-engine HP 57 Halifax. This in turn produced the HP 61 Halifax BIII, which had four Bristol Hercules engines. It is interesting to note that all the Royal Canadian Air Force squadrons based in Great Britain, at one time or another, flew the Halifax bomber. Towards the end of the Second World War, it became a heavy transport and glider tug aircraft.

Subsequently, during the Cold War era there was a requirement for a four-engine, nuclear bomb deterrent aircraft and Handley Page produced the Handley Page HP 80 Victor as part of the V-Bomber force, which remained in service for many years. A 1950s success was the HPR Handley Page (Reading) 7 Dart Herald turboprop airliner. The company also produced the successful Hastings and Hermes transport aircraft. Handley Page resisted amalgamations and mergers and in so doing hastened its demise. The two big aircraft manufacturers at the time were Hawker Siddeley and the British Aircraft Corporation. Its final product was the Handley Page Jetstream, a small turboprop, but this was not enough to sustain viability and the company folded in 1970.

Nord

Bourges, France
Founded 1 October 1954, became Aerospatiale 1 March 1970

After the Second World War the French aviation industries began to rebuild. Nord Aviation was created by the acquisition of SFECMAS (*la Société Française d'Etude et de Constructions de Matériel Aéronautiques Spéciaux*) by the SNCAN company. France's aviation industry rebuilt through acquisitions and mergers. Aircraft that were built by the company included general aviation, training, experimental (ramjet fighter), and utility transport. Just prior to Nord being created, the Noratlas was already on the books, being produced and in service with the French Air Force. It was probably one of their most successful aircraft, perhaps their very best. The Noratlas was a success at home with the Armee de l'Air and in the newly formed nation of West Germany. The German Air Force ordered just over 180 aircraft with 160 built by Flugzeugbau Nord company in Germany. The Noratlas won overseas orders with the Israeli, Portuguese, and Hellenic Air Forces. The Noratlas was also available for civilian customers. The SNECMA-manufactured Bristol Hercules 739/758/759 engines were installed in the Nord 2501 and 2502 models.

The company diversified and also built the Exocet sea-skimming, anti-ship missile. It became well known, infamous, for its role in the Falklands War where it sank a merchant ship and a Royal Navy destroyer. Four radio-controlled drones were completed in the 1950s. In the 1960s the Nord 260 and

262 high-wing turboprops were built. Aerospatiale would merge with the European Aerospace Corporation (EADS) in 2000 and finally become part of the Airbus Group.

Northrop

Hawthorne, California, US
Founded: 1939, became Northrop Grumman in 1994

The company was founded by Jack Northrop, who started his aviation career as a draftsman for the Loughead Aircraft Manufacturing Company, then the Douglas Aircraft Corporation, followed by an engineering position with the Lockheed Corporation where he led the team on the famous Lockheed Vega transport aircraft. He was determined to have his own company and after one of his companies was taken over by United Aircraft and Transport Company and one plant by a Division of Douglas Aircraft, he established his third company in Hawthorne. Here, Northrop focused on a flying wing project and production of the first American night-interceptor, the Northrop P61 Black Widow. The Bristol Hercules used the Northrop 8A-1 for test bed flying and the Gamma 2L.

Post-war, the company designed the F89 Scorpion Interceptor, the first American combat aircraft to have air-to-air guided missiles, including the nuclear unguided Genie rocket. In 1959 they designed the ground-launched SM62 Snark intercontinental cruise missile, capable of carrying a nuclear warhead. However, Northrop's most famous aircraft designed in the same era was the Northrop F5 supersonic light fighter. It focused on performance and low maintenance costs. For the next twenty-five years it was produced in many variants. Northrop's entry for the Light Weight Fighter program was the P600 (YF17 Cobra) which was later modified by McDonnell Douglas and became the F18 for the US Navy. Based on previous experience research, the B2 Spirit stealth bomber was produced. To remain competitive, Northrop bought the Grumman Aerospace Corporation.

The Hercules: The Other Engine That Helped Win the War

SARO

Saunders-Roe Limited
East Cowes, Isle of Wight
Founded: 1929, became Westland Aircraft in 1964

SARO was a British aeronautical and marine engineering company named after Alliott Verdun Roe, of AVRO fame, that took a controlling interest with J. Lord in the S. E. Saunders boat building company. Saunders had in 1926 built a prototype flying boat called the A4 Medina. SARO continued to build flying boats, and except for thirty-one London aircraft, they were not built in any significant quantities. In 1931, Spartan Aircraft Ltd of Southampton was merged into SARO. In 1938, a re-organisation took place with Saunders Shipyard Ltd and SARO Laminated Wood Products Ltd being formed. In 1938, the SARO Lerwick with Bristol Hercules II engines first flew. It was a medium-range, anti-submarine convoy escort and reconnaissance flying boat, built to replace the older biplane flying boats then in service.

During the Second World War, SARO built the Supermarine Walrus and Sea Otter under licence. At a separate location in Anglesey, Wales, the company modified and serviced Catalinas for the RAF. In 1947, SARO produced and flew the first jet-powered flying boat. In 1952, it produced the largest metal flying-boat ever produced, the double passenger deck Princess prototype. It had four coupled Bristol Proteus turboprop engines, a 66.9 m (219 ft 6 in) wingspan, and weighed in at 156,501 kg (345,025 lb).

The era of flying boats had passed and SARO diversified into autogiros, helicopters, and hovercraft (SRN 1). Westland Aircraft took over the rotary wing department in 1964. In its lifetime, SARO/Westland and its surviving components under various names were involved with lifeboats, power boats, and fast patrol boats. It proposed nineteen flying-boat designs, some of which went into production, others remained on the drawing board. In 1948, an Electronics Division was formed to produce computers and simulators. It had seven land-based projects including the

SR 53 mixed propulsion interceptor and jet and rocket engines. The Skeeter was a successful training and scout helicopter. In 1956, a hydrofoil was built for the Royal Canadian Navy. In 1958, the Black Knight research ballistic missile was flight tested in Australia at the Woomera Range.

Short (Shorts)

Short Brothers plc
Belfast, Northern Ireland
Founded 1908, became Bombardier 1989 and Spirit Aerosystems 2020

Longevity is associated with Shorts and its subsequent company names. It all started in 1897 when Eustace Short bought a coal gas filled balloon and, with his brother Oswald, decided to start a company to develop and manufacture balloons. By 1902, they were offering balloons for sale. They moved to a new factory next to the Battersea gasworks, very convenient! It took three years before they won a contract with the British Indian Army. Impressed with their work, Charles Rolls ordered his own personal balloon to compete in the 1906 Gordon Bennett Cup balloon race. In 1908, upon hearing reports from the Aero Club of Great Britain about the Wright Brothers' demonstration at Le Mans, France, Oswald made the decision that ballooning was finished and the aeroplane would now take over the aviation scene. They persuaded their brother Horace to join them in their new aircraft venture. Shorts immediately got two orders, one for a glider from Charles Rolls, and the other for an aircraft from Francis McClean. By 1909, the first aircraft, Short No 1 biplane, was exhibited as an airframe, no covering, and Shorts had obtained the rights to build copies of the Wright's design. The factory was built on the Isle of Sheppey and work began on six aircraft. Shorts became the first manufacturing company to have volume production of a design.

In 1911, Shorts began building a long series of naval aeroplanes. The Short Admiralty Type 184 was the most successful, over 900 built, a folding-wing bomber and torpedo carrying seaplane. A land version, the Short Bomber, was used by the Royal Flying Corps. At Cardington, Bedfordshire, balloons and dirigibles were produced. Post-war and into the 1930s Shorts focused on the production of flying boats. Singapore, Calcutta, and Sunderland aircraft followed to meet the requirement for long-distance flights, including trans-Atlantic flights, and military patrol aircraft.

In 1936 the Air Ministry opened facilities in Belfast, Northern Ireland. It was a cooperative venture called Short & Harland Ltd. The Hercules-powered Short flying boats were the S26 G Class, S45 Seaford, and S45 Solent. The Short Stirling was a four-engine, heavy bomber. Other names of note with different engines were the Shetland, Hythe, and Sandringham. In 1947, all the facilities were consolidated in Belfast. Short-haul and heavy lift freighters were built. A regional jet was planned but was cancelled when the company was bought by Bombardier, which had its own CRJ100. The company, after having some 120 or more projects, not all built, had diversified into airships, missiles, rotorcraft, UAVs, and drones.

Vickers

Vickers Engineering
(Sheffield), London
Founded: 1828, became British Aircraft Corporation 1960 and BAE Systems 1999

Another company with longevity was Vickers Engineering, just a few years younger than Shorts. It all started as a steel foundry in Sheffield run by Edward Vickers and his father-in-law George Naylor. The company gained fame and recognition for casting church bells. Edward was joined by two brothers,

The Hercules and Aircraft Companies

Tom and Albert, and the family team started to build the casting and forging company. In 1911, it was known as Vickers Ltd and expanded into the aviation business with Vickers Ltd (Aviation Department) and unusually, the Vickers School of Flying at Brooklands, near Weybridge, Surrey. This was one of Britain's first banked race tracks and aerodromes. In 1927, Vickers merged with Armstrong Whitworth to become Vickers-Armstrongs and in 1928, Vickers (Aircraft) Ltd. In 1938, both companies were organised as Vickers-Armstrongs (Aircraft) Ltd. The names Supermarine and Vickers were still retained for their products. In 1960, Vickers became part of the British Aircraft Corporation along with Bristol, English Electric, and Hunting Aircraft. Canadian Vickers was a subsidiary from 1911 to 1944 when it was taken over by Canadair. As further described in Chapter 6, The Vickers Wellesley (1935 test bed), Wellington (1936), Viking (1945), Valetta (1947), and Varsity (1949) were all powered by Bristol Hercules engines.

A mention should be made of four other aircraft among the nearly 200 designs, some built, some not: the Royal Flying Corps heavy bomber Vickers Vimy that crossed the Atlantic Ocean non-stop in 1919; the 1948 medium-range turboprop airliner, the Vickers Viscount; the 1955 Vickers Valiant high-altitude, jet bomber, part of the nuclear weapon carrying 'V bomber' RAF strategic deterrent force; and the 1959 Vickers Vanguard, which was quickly overshadowed by the emerging jet-powered airliners.

6

THE HERCULES POWERED AIRCRAFT

To reiterate: there were so many variants and sub-variants that it is impossible to verify statistics definitively unless they are attributed to an actual aircraft constructor number.

BRITISH AIRCRAFT

The various marks of Hercules engines were chosen by these manufacturers to power the following listed aeroplanes. Sometimes it was the entire production line, the Armstrong Whitworth Albemarle, or sometimes just one prototype aeroplane, the Avro York C.II. The following are general aeroplane descriptions, including the series of Hercules engine installed in specific models.

The engine installations are predominately in bomber or transport type aeroplanes but include maritime patrol flying boats and one fighter aeroplane. Ten British manufacturers listed alphabetically here used the Hercules in twenty-four different models of aeroplanes. These vary from the Bristol Beaufighter in 1939 to the Bristol Superfreighter in 1953. Five foreign manufacturers, listed alphabetically, used the Hercules in six different aeroplanes, some only for test purposes. The aeroplane numbers listed are totals produced, unless the number of Hercules-engined models is known.

Where the Hercules engine was only used in a prototype or development model, it was not possible to be sure of the specifications for that particular model. Instead, the developed model with other engines, the Blackburn Beverley with the Centaurus engine being an example, is listed to show how the Hercules engine contributed to the initial development of that particular aeroplane and then how it progressed to its final model.

The requirement for engine power grew as the weight of the aeroplanes and the desire to increase speed grew. The design of the bomber aeroplanes of the Second World War focused on the requirement to carry a greater bomb load both faster and higher to the target. The Hercules engine responded by increasing its horsepower from 1,150 hp in the Hercules I to 1,725 hp in the Hercules XIX. That is an increase of sixty-seven per cent, no small feat. Post-war, the horsepower reached 2,080 hp with the Hercules 762 in the Handley Page Hermes.

Another success story was the reliability of the Hercules engine as it opened up the world with long-range flights by flying boats. These scheduled flights included such routes as multiple daily flights from England to South Africa and non-stop transatlantic flights from Ireland to Newfoundland, which then continued on to the US. The size of some of these flying boats, two decks and twenty-seven metres (eighty-eight feet) long, allowed the public, the rich public, to travel in style.

Hercules/Merlin Development Comparison

	Hercules	Year		Merlin
I	858 kw (1,150 hp)	1936		
		1937	I	768 kw (1,030 hp)
II	1,025 kw (1,375 hp)	1938		

The Hercules: The Other Engine That Helped Win the War

Hercules		Year	Merlin	
III	1,044 kw (1,400 hp)	1939	III	768 kw (1,030 hp)
		1940	XIX	1,037 kw (1,390 hp)
XI	1,186 kw (1,590 hp)	1941	31	1,037 kw (1,390 hp)
XVI	1,286 kw (1,725 hp)	1942	61	955 kw (1,280 hp)
XIX	1,286 kw (1,725 hp)	1943	67	981 kw (1,315 hp)
100 Civil	1,249 kw (1,675 hp)	1944	102 Civil	1,219 kw (1,635 hp)

Armstrong Whitworth Albemarle

First Flight: 20 March 1940
Built by: A. W. Hawksley Ltd, Gloucestershire. Production 1941–44: 602
Operational 1943–1946

The Albemarle was a Second World War twin-engined transport aircraft. It was originally designed as a medium bomber, but subsequently was favoured as an aerial reconnaissance and transport aircraft requiring extensive redesign mid-development. This delayed its introduction to No. 295 Squadron RAF until January 1943. The Albermarle was now used for transport duties, paratrooper transport, and glider towing. Albemarle Squadrons did participate in the Normandy Landings and the assault on Arnhem during Operation *Market Garden*. In October 1942, Russia received a few Albemarles.

Development: The aeroplane was designed using wood and metal construction without the use of any light alloys, allowing less experienced non-aviation companies to build it. Armstrong

Whitworth used a tricycle landing gear and Rolls-Royce Merlin engines but could have 'shadow factory' Bristol Hercules engines installed. The change in focus allowed the Albemarle to add more fuel and dorsal and ventral retracting guns.

Production: In 1938 200 aircraft were ordered 'off the drawing board'. No prototypes were constructed. The first aircraft was assembled and flown at Hamble Aerodrome by Air Service Training. The first test flights have been described as ordinary and free of flaws. The production run was by A.W. Hawksley of Gloucester and over a thousand sub-contractors completed individual parts and sub-assemblies MG Motors produced the forward fuselage for example, Rover constructed the wing centre section, and Harris Lebus produced the tailplane units. Production ceased in 1944.

Design: The Albemarle was a mid-wing cantilever monoplane with twin fins and rudders. It started as the Bristol 155, a derivative of the Bristol Beaufort, but was transformed to an Armstrong upon the death of its designer, Frank Barnwell, in a flying accident. The fuselage was built in three sections, unstressed plywood over a steel frame. The damaged section could be replaced. The Frise type ailerons and tailplane were also wood. The Lockheed-designed landing gear hydraulic system retracted the nose gear backwards into the front fuselage and the main gear into the engine nacelles. Two Bristol Hercules XI engines drove a three-bladed, De Havilland Hydromatic propeller unit. The two pilots sat side by side in the forward portion of the cockpit, the navigator was in the nose. The bomb-aimer was in the crew hatch in the underside of the nose. The dorsal Boulton-Paul turret had four machine guns. Some models had a ventral turret. Rarely used as a bomber, later types of the Albemarle became 'GT' general transport or 'ST' special transport. Modifications were also incorporated for towing gliders and ferrying paratroopers.

Operational history: The Soviet Air Force ordered 200 aircraft in October 1942. No. 305 FTU at RAF Erroll near Dundee,

The Hercules: The Other Engine That Helped Win the War

Scotland, was set up to train the Russian crews. The first operational RAF operation was to drop leaflets over Lisieux in Normandy, France on 9 February 1943. In 1943, the Albemarles were involved in the Sicily campaign. However, the height of the aircraft's career was during the Normandy D-Day campaign. The squadrons were towing the Airspeed Horsa, Waco Hadrian, and Hamilcar gliders.

Variants: Eight variants of the Hercules-powered Albemarle were built, with two, prototypes only, using a Rolls-Royce Merlin and Wright Double Cyclone engines.

Operators Military:
Soviet Air Force. RAF: Squadron (7), Flight (2), Operational Training Unit (2), Heavy Glider Conversion Unit (3), Glider Training School (1), Ferry Training Unit (2), Torpedo Development Unit (1), Telecommunication Unit (1), Airborne Forces Experimental Establishment (1), Coastal Command Development Unit (1), Central Gunnery School (1), Bomber Development Unit (1), Operation Refresher Training Unit (1).

Operators Civil:
Royal Aircraft Establishment, Aircraft & Aeronautical Experimental Establishment (A&AEE), and De Havilland Propellers.

Albermarle ST Mk 1
Crew: Four, two pilots, navigator, and bomb-aimer (Transport)
Six, two pilots, navigator/bomb-aimer, radio operator and two gunners (Bomber)
Capacity: Ten troops
Max takeoff weight: 16,556 kg (36,500 lb)
Fuel capacity:3,500 L (769 imp gal) normal, 6,360 L (1,339 imp gal) with auxiliary tanks
Powerplant: Two Bristol Hercules XI engines, 1,190 kw (1,590 hp)
Propellers: 3-bladed de Havilland Hydromatic

Performance
- Maximum speed: 426 kmh (265 mph) at 3,200 m (10,500ft)
- Cruise speed: 270 kmh (170 mph)
- Range: 2,100 km (1,300 miles)
- Service ceiling: 5,500 m (18,000 ft)

Armament
Guns: Four .303 in (7.7 mm) Browning machine guns in dorsal turret, two .303 in (7.7 mm) machine guns in ventral turret (first prototype only)
Bombs: Internal 2,000 kg (4,500 lb)

Avro Lancaster

First Flight: 09 January 1941
Built by: Armstrong Whitworth, B.II Production: 300, All Models: 7,377
Operational: February 1942

The Avro Lancaster was a four-engined heavy bomber built by Avro to meet a specification issued by the Air Ministry. The Handley Page Halifax was also developed at the same time to meet the same criteria. Along with the Short Stirling, these two bombers would form the core of the RAF heavy bomber complement. The Lancaster had its origins in the two-engined Avro Manchester of similar fuselage design. The Manchester was designed to deliver a torpedo carried internally and make shallow dive bomb attacks. Roy Chadwick, the Avro designer, designed the aircraft primarily for the Rolls-Royce Merlin engine. However, to avoid a possible future shortage of Merlin engines the BII version of the Lancaster was designed using the Bristol Hercules engine. 300 aircraft of this version were built, representing approximately four per cent of the more than 7,300 produced. It first saw service with RAF Bomber

The Hercules: The Other Engine That Helped Win the War

Command in 1942. The strategic bomber offensive, under the leadership of Air Marshal Sir Arthur 'Bomber' Harris, gained momentum and the Lancaster played a leading role in the night-time campaign. It served also with Commonwealth and European squadrons associated with the RAF. The unique unobstructed long bomb bay enabled the Lancaster to carry outsized and speciality bombs. The 'Lanc", as it was commonly called, was one of the most heavily used night-bombers, delivering over 600 million kilograms (over 608,000 long tons) of various sizes and types of bombs. During the latter stages of the war, the Lancaster became a test bed for the developing turbojet engines. Post-war it would be developed into the Avro Lincoln and the Avro Lancastrian civil airliner.

Development: In the early 1930s, before the threat of another war, the RAF was primarily interested in developing the two-engine bomber. This was to reduce the demands on an aviation industry trying to cope with the introduction of many types of aircraft for future consideration. The engine manufacturers were encouraged to develop larger and more powerful engines to improve the performance. Meanwhile, the United States had taken the opposite tack of using four engines to deliver the required performance in range and power. By 1936, the RAF would change its view because of the worsening political situation in Germany. In the meantime, Avro submitted and constructed a two-engine medium-bomber called the Avro Manchester. It entered service in 1940 with Rolls-Royce Vulture engines, which proved unreliable, and it was withdrawn from service in 1942. By 1940, Chadwick was working on a four-engine version with the less powerful, but more reliable, Rolls-Royce Merlin engines. Initially called the Type 683 Manchester III, it was subsequently renamed Lancaster. It was basically a Manchester with a new wing to accommodate the four engines. It first flew in January 1941. The centre fin was removed and the remaining elliptical fins were increased in size. The success of the Lancaster caused the remaining Manchesters on the assembly line to be changed to Lancaster BIs and by October 1941 the production Lancaster had its first flight.

Production: The first order of 1,070 aircraft would become the first of many orders culminating in a total of over 7,300. Avro built the Lancaster at Chadderton, Lancashire, but the company soon realised that it alone would be unable to cope with the wartime demand. A 'Lancaster Aircraft Group' was created to manufacture complete aircraft units or supply components, major or minor, to the assembly companies. Major companies involved were Armstrong-Whitworth and Metropolitan-Vickers. On a lesser scale were Austin Motors, Vickers Armstrong, and Victory Aircraft in Canada. The Canadian models had American Packard-built Merlin engines and American style instruments and electrical systems. The basic Lancaster remained the primary model with minor alterations to house specialised equipment and outsized or special bombs. The propellers would change in time from the 'needle nose' to 'paddle' type. The engines would be either Rolls-Royce or Packard built Merlins. The exception would be the 300 Bristol Hercules versions.

Design: The mid-wing cantilever monoplane is constructed using five wing sections and five fuselage sections. The ten sections were manufactured separately and were completed with the required equipment before assembly to the other completed sections in an effort to increase production. The oval-shaped fuselage, specially designed for a maximum structural strength to weight ratio, had an all-metal covering. The ailerons were an exception, being fabric covered. The hydraulically operated landing gear retracted rear-wise into recesses in the inner engine nacelles. The tailwheel did not retract. The tail unit had a two, distinctive elliptical fin rudder configuration.

It had favourable flying characteristics but was not without challenges. In a dive the longitudinal stability at high speeds tended to steepen the angle. The bomb aimer lay prone on the floor to operate the bomb sight or stood up to operate the nose

The Hercules: The Other Engine That Helped Win the War

gun. The escape hatch was also in this nose compartment and was difficult to exit wearing a parachute. Perhaps a black mark for the Lancaster, as it did not have a good record for crews exiting by this hatch. The single pilot and flight engineer, using the 'Dicky seat', sat under the large canopy. The navigator and wireless operator sat in their own curtained-off section from the flight deck compartment. Behind them were the two main wing spars that passed through the fuselage, creating a crew obstacle. The mid-upper gunner sat on a canvas sling. The fuselage door was further aft on the starboard side along with the Elsan chemical toilet. The tail gunner entered his gun by a small hatch in the very rear of the fuselage. Being exposed to the elements, the gunners wore electrically heated suits in the latter stages of the war.

Armament: Initially the Lancaster had four (Nash & Thompson) Frazer Nash hydraulically operated turrets. The ventral turret was soon removed for weight and effectiveness reasons. Boulton Paul turrets were sometimes used in the mid-upper position. Eight 7.7 mm (.303 in) Browning machine guns, two in nose, two in dorsal, and four in rear turret, were installed. The unobstructed 10 m (33 ft) long bomb bay separated the Lancaster from the other bomber aircraft. It could carry a large variety of types and weights of bombs with only modification to its bomb bay doors. For example, the 'Cookie' bombs varied from 1,800 kg (4,000 lb) to 5,400 kg (12,000 lb) bombs. Other types included the Small Bomb Container, incendiary, armour piercing, the famous Dam Buster bomb, and the 4,200 kg (9,250 lb) 'Upkeep' bouncing bomb. To finish off the Lancaster's incredible inventory capability it carried the 5,400 kg (12,000 lb) 'Tallboy' and 10,000 kg (22,000 lb) 'Grand Slam' bomb. Part of its armament were four different bombsights and ten radio, radar, and countermeasure systems.

Operational history: It is beyond the scope of this book to recount the contribution of the Lancaster to winning the Second World

War. Suffice to say that many books have been written about this famous aircraft and its exploits. One of them, published by Amberley Books, is my own, *The Lancaster*. After the war it repatriated troops, operated as a freighter, developed inflight refuelling, took part in the Berlin Airlift, and was converted to a civil airliner.

Variants

Model	Description
BI	Original Rolls-Royce Merlin XX engines, later Merlin 22, 24.
BI Special	Aircraft adapted to carry Tallboy, bomb bay doors bulged, and Grand Slam, bomb bay doors removed.
PR1	B1 modified for photographic reconnaissance, Nos 82, 541 RAF Squadrons.
BI (FE)	Tropicalised version for service with Tiger Force in the Far East
BII	Bristol Hercules VI or XVI engines. The VI engines had manual mixture control requiring an extra lever on the throttle quadrant.
BIII	Packard built Merlin engines with Bendix-Stromberg pressure-injection carburettors.
BIII (Special)	Modified aircraft to carry the dam busting *Upkeep* bouncing bomb.
ASRIII/ASR3	BIII aircraft modified for air-sea rescue operation.
GR3/MR3	BIII modified for maritime reconnaissance.
BIV	Larger version with Rolls-Royce Merlin 85/68 engines. Renamed Lincoln B1.
BV	As above with Rolls-Royce Merlin 85 engines. Renamed Lincoln B2.
BVI	Converted BIIIs with Rolls-Royce Merlin 85/87 engines. High altitude version withdrawn from service in 1944 due engine vibration problems.
BVII	Final version.
BX	Canadian built BIII.
XPP	Lancastrian for Trans Canada Airlines

The Hercules: The Other Engine That Helped Win the War

Operators Military (Squadrons): Argentine: Argentine Air Force (1). Australia: Royal Australian Air Force (3). Canada: Royal Canadian Air Force (14), Rescue (1), OTU (1), Reconnaissance (3), Photography (2). Egypt: Royal Egyptian Air Force. France: Aeronavale (5)

Poland: Polish Air Force in exile (1). Soviet Union: Soviet Naval Aviation (repaired crashed aircraft). Sweden: Swedish Air Force (one test aircraft). United Kingdom: Fleet Air Arm (1), RAF (66), Heavy Conversion Unit (16), Lancaster Finishing School (3), OTU (1), Conversion Unit (2)

Operators Civil:
Argentine (1), Canada (3), United Kingdom (5)

Avro Lancaster BI
Crew: (Seven) Pilot, flight engineer, navigator, bomb aimer/nose gunner, wireless operator, mid-upper gunner, rear gunner.
Max takeoff weight: 30,844 kg (68,000 lb)
Fuel capacity: 9,792 L (2,154 imp gal) normal; 13,429 L (2,954 imp gal) with auxiliary tanks
Powerplant: Four Rolls-Royce Merlin XX 950 kw (1,280 hp)
Propellers: 3-bladed De Havilland Hydromatic.

Performance
- Maximum speed: 454 kmh (282 mph) at 3,962 m 13,000(ft)
- Cruise speed: 320 kmh (200 mph)
- Range: 4,070 km (2,530 miles)
- Service ceiling: 6,500 m (21,400 ft)

Armament
Guns: Eight 7.7 mm (.303 in) Browning machine guns, two in nose, two in dorsal, and rear turret. Two in ventral position, early aircraft only.
Bombs: Internal 6,400 kg (14,000 lb) normal load.

Avro 689 Tudor

First Flight: 14 June 1945
Built by: Avro, Production VII: 1, Total: 38
VII: prototype only
Introduction: 1947

The Tudor was an airliner based on the Avro four-engined Lincoln bomber, which was descended from the Lancaster bomber. It was Britain's first pressurised airliner. The tail wheel landing gear configuration dated the Tudor and customers were more likely to buy American products with tricycle landing gear.

Development: Avro began work in 1943 on a commercial adaptation of the Lancaster IV bomber. It was later renamed the Lincoln. The Brabazon Committee issued specifications for nine post-war commercial aircraft. Avro proposed a circular section pressurised Lincoln fuselage and wing with a single fin and rudder, which retained the Rolls-Royce Merlin 102 1,305 kw (1,750 hp) engines.

Design: The Tudor was a low-wing cantilever monoplane with four engines and a retractable tail wheel. The wing was a five-piece all-metal twin-spar construction. The untampered centre section carried the inboard engines and the main landing gear. Trim and balance tabs were fitted to the ailerons. The hydraulic split flaps were in three sections located on the trailing edge of the inner and centre wing sections. Fuel was carried in bag tanks in the fuselage centre section and inner wings. The control surfaces were mass balanced and had controllable trim and servo tabs. The main landing gear retracted into the engine nacelle and the twin tailwheels retracted into the fuselage enclosed by doors.

Airline history: The Tudor I had been planned to fly the North Atlantic route, but due to instability problems and American competition from the Lockheed Constellation and Boeing Stratocruiser, it did not happen. Twelve of the twelve-passenger aeroplanes were built. An enlarged sixty-seat version, Tudor II, was designed for the Commonwealth routes but suffered loss of performance during high and hot conditions. The prototype crashed in August 1947 killing the famous Avro designer Roy Chadwick. The second prototype was converted to the Tudor VII with Hercules engines. The government was criticised for its policy of development which led to aircraft which were not competitive being overpriced and lacking in performance. Some variants were used with limited success by British Overseas Airways Corporation (freight), but BOAC cancelled its passenger orders and bought the Canadair North Star instead. British South American Airways had short, limited success on multi-stop transatlantic flights from England to South America. However, two aeroplanes disappeared over the Atlantic causing the grounding of the fleet and conversion to unpressurised freighters. The Tudor did, however, contribute to the Berlin Airlift as a fuel tanker.

Variants: There were thirteen variants including prototypes and a Rolls-Royce Nene test bed.

Operators Civil: BOAC, BSAA, and later by charter and freight airlines.

Avro 689 Tudor VII
Crew: (Four) Two pilots, flight engineer, radio operator, navigator
Capacity: unknown. Previous versions had seat/berths combinations
Max takeoff weight: 34,473 kg (76,000 lb)
Fuel capacity: 15,000 L (3,300 imp gal)
Powerplant: Four Bristol Hercules 120 engines; 1,305 kw (1,750 hp)
Propellers: 4-bladed Rotol, constant-speed, fully-feathering.

Performance
- Maximum speed: 557 kmh (346 mph) at 6,096 m (20,000 ft)
- Cruise speed: 483 kmh (300 mph) at 6,858 m (22,500 ft)
- Range: 7,500 km (4,660 miles) with maximum fuel
- Service ceiling: 9,200 m (30,100 ft)

Avro York

First Flight: 05 July 1942
Built by: Avro, C.II Production: 1, Hercules XVI engines. Total: 254 + 4 prototypes
Introduction: 1944

The Avro York was a British transport aircraft developed by Avro during the Second World War. It borrowed heavily from other Avro aircraft on the drawing board and in production. Mid-point in the war the government and Avro foresaw the need for military and civilian transports.

Development: In a very uncertain time of the Second World War with the emphasis on bomber aircraft, Avro simultaneously, bravely, embarked on the development of a new civil-oriented transport aircraft. There was certainly a need and BOAC had been formed in 1940 to service all the nation's overseas routes. But how big would be the future demand for civil transports be? A risky venture indeed, unless some parts of the aircraft existed already. It did, in the form of the Avro Lancaster heavy bomber.

Roy Chadwick took the basic design of the Lancaster and adopted the fuselage for passenger use. He called the private venture the Type 685 and to expedite its development used the Lancaster's wings, tail assembly, landing gear, and Rolls-Royce Merlin engines with a square-section fuselage to provide sufficient space for passengers. Within eighteen months of the Lancaster, the York took to the air on 5 July 1942 at Ringway Airport, Manchester. It required a central extra fin to maintain control and directional stability.

In response to successful flight testing at RAF Boscombe Down, the Air Ministry ordered three more prototypes to be built. The first prototype was rebuilt to CII standards and was the model used for the installation of the Bristol Hercules VI radial engines. However, it was later decided to standardise on the Merlin engine, so this prototype, LV626, was the sole Hercules-powered York in existence.

The Hercules: The Other Engine That Helped Win the War

Production: The Avro focus, and the war's, was on the production of the Lancaster bomber. This would mean that the York would be on a limited production schedule. By the end of 1943, only seven aircraft had been built, four prototypes and three production aircraft. However, in spite of wartime conditions, three aircraft per month were scheduled for 1944. In March 1943, RAF Transport Command was created and that formalised the requirement for transport military aircraft. The York was the first British aircraft to be used in quantity by the Command. The first RAF production order was for 200 aircraft The majority were passenger versions with some batches of freighter and combined passenger-freighter versions. Some early RAF aircraft were diverted to BOAC for their overseas service. Initial assembly and testing of production Yorks was at Ringway, Manchester, and later at Yeadon, Leeds, and Woodford, Cheshire. A single pattern aircraft was constructed by Victory Aircraft 'shadow factory' in Malton (Toronto), Canada before war's end. It was to be a thirty-aircraft production order.

Design: The Avro York was a high-wing, cantilever, all-metal monoplane derived from the Avro Lancaster bomber. The semi-monocoque fuselage had a flush-riveted skin and was built in five main sections. It had seven internal fuel tanks housed between the two-spar construction of the wings. The tapered edges of the outboard wing panels had detachable wing tips. The wings had all-metal, hydraulically operated, split trailing edge flaps. The four Rolls-Royce Merlin engines were mounted in underslung nacelles attached to the front spar. The propellers were metal de Havilland Hydromatics.

A twenty-one seat, three-abreast arrangement split between two cabins was a typical seating arrangement. The main entrance was between the two cabins along with the cloakroom and lavatory. The galley and baggage area were at the rear of the cabin. An interesting feature, remembering that it was an early 1940s design, was that the emergency exits were in the roof of each cabin. The passengers endured a very noisy ride due to the Merlin engines.

The Hercules Powered Aircraft

Airline/Operational history: Civilian: (April 1944) BOAC inaugurated service, England-Morocco-Cairo, Egypt, and further operations, with twenty-five York aircraft, in collaboration with No.216 Group RAF. In partnership with South African Airways, service was established to Johannesburg. These aircraft were specially outfitted with twelve sleeping berths in addition to passenger seats. Post-war, BOAC expanded their service, taking over from Shorts flying boats on such routes as Cairo to Durban. British South American Airways Company used the York on their Caribbean and South America routes. By October 1952 BOAC retired the York from passenger service and established a freight operation with the aircraft. Five years later, the aircraft were passed on to several independent British airlines. The largest operator of Yorks was Skyways, whose routes included long distance troop flights to Jamaica. The end of York service was in 1964, a successful twenty years for both the aircraft and the Rolls-Royce Merlin engine.

Military: (1945) No. 511 Squadron RAF was the first squadron to be fully equipped with the Avro York. Eventually, ten RAF squadrons would have them on strength. The RAF used it on all their trunk routes such as the critical England-India route. The Yorks flew many sorties during the Berlin Airlift of 1948–1949. The York was the most effective British transport and alone carried close to half the tonnage of the British contribution.

VIP Service: One of the prototypes, LV633, was outfitted as the personal transport and flying conference room for then Prime Minister Winston Churchill. (1945) MW140 was the personal transport for the Duke of Gloucester, Australia's Governor-General. It was used in 1945 to repatriate prisoners-of-war from Singapore. It was returned to the Air Ministry in 1947. MW102 was the flying office for the Viceroy of India, Lord Mountbatten.

Variants: Four Avro 685 prototypes, one CII Bristol Hercules prototype, York I civilian transport, York CI military transport, York C3 (Victory Aircraft).

Operators Military: Australia (Governor-General's Flight RAAF), France, South Africa, RAF, TRE, A&AEE, RAE, AFEE

Operators Civil: Aden, Argentina, Canada, Iran, Lebanon, South Africa, Britain (Air Charter, BOAC, British South American Airways, Dan-Air, Eagle Aviation, Hunting-Clan Air Transport, Scottish Airlines, Skyways, Surrey Flying Services)

Avro York
Crew: (Five) two pilots, navigator, wireless operator, cabin steward
Max takeoff weight: 29,484 kg (65,000 lb)
Fuel capacity: 11,270 L (2,976 imp gal)
Powerplant: Four Rolls-Royce 24 engines 980 kw (1,280 hp)
Propellers: 3-bladed, fully feathering, constant speed propellers

Performance
- Maximum speed: 480 kmh (298 mph)
- Cruise speed: 375 kmh (233 mph)
- Range: 4,800 km (3,000 miles)
- Service ceiling: 7,000 m (23,000 ft)

Blackburn & G.A.L. 60 Universal Freighter 1

First Flight: 20 June 1950
Built by: General Aircraft, Production: 2, Total: 48

The General Aircraft designed and built the G.A.L. Universal Freighter 1 which was the basis of the Blackburn Beverley heavy transport which served with the Royal Air Force from 1957 until 1967. Blackburn was essentially a builder of naval fighters. This was the only land-based transport aeroplane that the company ever built.

Development: The first aircraft was dismantled at the Hanworth site near Feltham, Middlesex, and moved by road to Brough Aerodrome,

The Hercules Powered Aircraft

East Riding of Yorkshire. General Aircraft realised that they did not have the room to develop such a large aircraft and in 1949 merged with Blackburn Aircraft Ltd. Blackburn would go on to build the Blackburn Buccaneer maritime strike aircraft at Brough. The second aircraft, G.A.L.65, was modified with clamshell doors replacing the original ramp and door, seating for thirty-five passengers in the tailplane boom, and replacing the original Bristol Hercules engines with Bristol Centaurus engines. It went on to a successful career with Centaurus engines. The RAF originally ordered twenty aircraft in October 1952, now known as the Blackburn Beverley C.1 (Cargo Mark 1), and further orders would bring the total to forty-seven.

Design: The Beverley is an unusual design for the times, with fixed landing gear and an atypical seating arrangement. It is a high-wing monoplane with a large fuselage (170 cu m, 6,000 cu ft) and a tailboom supporting a twin fin tailplane. The tailboom had seating for thirty-six and allowed access to the fuselage underneath via unique removable clamshell doors. The Beverley could accommodate bulky loads and was designed to operate on unprepared short runways. In 1957, it was the largest aircraft in the RAF. The paratroopers in the main cargo area exited via side doors and those in the tailboom through an exit in the floor just forward of the tailplane.

Operational history: It saw service with squadrons in Kenya, Aden, Singapore, and Bahrain. The author saw this 'aluminium cloud' approaching Guernsey Airport while at a scout camp in the early 1960s.

Variants: General Aircraft built two prototypes before merging with Blackburn, who built forty-seven B-101 models.

Operators Military (Squadrons): RAF (6), Operational Conversion Unit.

B-101 (Centaurus engine version)
Crew: (Four) Two pilots, radio operator, navigator.

Capacity: 94 troops or 70 paratroopers with additional seating in the tailboom.
Max takeoff weight: 61,235 kg (135,000 lb)
Fuel capacity: 31,300 L (6,880 imp gal)
Powerplant: Four Bristol Centaurus 173 engines; 2,130 kw (2,850 hp)
Propellers: 4-bladed de Havilland reversible-pitch

Performance
- Maximum speed: 383 kmh (238 mph)
- Cruise speed: 278 kmh (173 mph)
- Range: 2,100 km (1,300 miles) at 2,400 m (8,000 ft with 13,000 kg (29,000 lb) payload
- Service ceiling: 4,900 m (16,000 ft)
- Takeoff distance to 15 m (50 ft): 408 m (1,340 ft)
- Landing distance from 15 m (50 ft): 277 m (910 ft)

Bristol Beaufighter

First Flight: 17 July 1939
Built by: Bristol Aeroplane Company, Production: 5,928
Operational 27 July 1940

Nicknamed the 'Beau', the Bristol Beaufighter evolved from the Bristol Blenheim light bomber to the Bristol Beaufort torpedo-bomber to a multi-role aircraft used extensively by the RAF and Fleet Air Arm. It proved itself to be a capable night fighter with airborne interception radar, a rocket-armed ground attack aircraft, and as 'Torbeau', a maritime torpedo anti-shipping strike aircraft. The British aircraft was also modified slightly by the Australian Department of Aircraft Production and served with distinction in the Battle of the Bismarck Sea.

Development: The Beaufighter started its life as a private venture by Bristol to enhance the Beaufort to a cannon-armed fighter derivative. The modest performance would be improved by the new Bristol Hercules engine. While there was some scepticism at the Air Ministry, who thought that the design was too big for a fighter. It benefitted from the delay of the Westland Whirlwind fighter. War was on the horizon; time was of the essence. The Beaufighter was a modified, slim version of the Beaufort. Before the prototype's first flight, an order for 300 aircraft was issued. Then began a war of engines, including a Rolls-Royce Griffon and Merlin proposals. The Merlin version, with over 300 built, was underpowered and had control issues during takeoff and landing. Various combinations of machine guns, cannons, and rockets ensued. Lord Beaverbrook, Minister of Aircraft Production, was very keen to get the aircraft into service as a night fighter with the capacity to incorporate airborne radar equipment. By sub-contracting components and the addition of shadow factories, the rate of production would greatly improve. The designation of F for Fighter Command and C for Coastal Command were used to distinguish versions.

Design: The two-seat twin-engine mid-wing cantilever all-metal aircraft was designed to be a long-range day and night fighter. The monocoque construction comprised three sections, which used 'Z-section' frames and 'L-section' longerons to join and reinforce the sections. The wing consisted of two main spars with a stressed-skin covering. The ailerons were metal-framed with fabric covering. The flaps were hydraulically operated along with the main landing gear. The brakes were pneumatically operated. The Bristol Taurus engines of the Beaufort were replaced by the more powerful series of two-speed, supercharged Hercules engines. The propellers were three-bladed Rotol, constant-speed, fully feathering, initially with wooden and then metal blades. Initial problems with vibration and centre of gravity issues gave the Beaufighter its characteristic stubby appearance. The crew accommodation had the pilot in a fighter type cockpit and the navigator/radar operator sat under a small Perspex bubble. Both crew had their own escape hatches in the floor of the aircraft, the pilot's being more difficult to reach as it was behind his seat! The armament was designed to be carried in the lower fuselage and wings. A small bomb load could be carried externally, and four Hispano cannons were mounted in the lower fuselage area. The cannons were supplemented with Browning machine guns in the wings. The recoil was extremely strong when firing cannons and guns together, causing a reduction in airspeed. The spacious fuselage allowed the mounting of airborne interception radar in the nose and along with the firepower was a formidable combination for the early stages of the war.

Hercules	Bristol Type 156
II	Beaufighter IF
III	Beaufighter I
VI	Beaufighter I, VIC, VIF, XTF
X	Beaufighter I
XI	Beaufighter I
XVI	Beaufighter VIC, VIF, TFX
XVII	Beaufighter TFX, XIC
XVIII	Beaufighter TFX, XIC, 21
130	Beaufighter VIC

Operators Military (Squadrons): Royal Australian Air Force (7). Royal Canadian Air Force (4).

Dominican Republic, ten VIF aircraft. Israel, four TF.X aircraft. New Zealand Squadrons, RAF (2). Polish Air Force, in exile Great Britain, (1). Portuguese Navy, fifteen TF.X aircraft. South African Air Force (2). Turkish Air Force, thirty-two approximately TF.X. United Kingdom, RAF (59), one flight, Fleet Air Arm (15). United States Army Air Force (4).

Operational history: Initially, the slow and heavy Beaufighter was introduced to squadron service without airborne radar. These squadrons were converted from the Blenheim light bomber. It was not until October 1940 that the first enemy aircraft destroyed was reported. The next month was the first kill by a radar-equipped night fighter version. By May 1941, advanced radar was installed and the Beaufighter was being credited with fifty per cent of enemy aircraft shot down. A year later the faster De Havilland Mosquito took over as the main night fighter. However, the rugged and reliable 'Beau' would contribute to the war effort in other ways. One famous raid was the dropping of the 'tricolore' on the Arc de Triomphe. It saw service overseas in 1943 by the United States Army Air Force in the Mediterranean. The Beaufighter was especially suited to Coastal Command as a long-range fighter and had a very successful operation based in Malta against enemy shipping and ground targets. The Beaufighter continued development with torpedo and rocket versions to enhance low-level attack capabilities. It was deployed to India for operations in Burma and Thailand. The Royal Australian Air Force used it very successfully in the Southwest Pacific theatre in New Guinea.

Bristol Type 156 Beaufighter Variants

Mk	Description
Mk IF	Night fighter, AI Mark IV radar
Mk IC	Coastal Command
Mk IIF	Rolls-Royce Merlin XX engine
Mk III/IV	Bristol Hercules and Rolls-Royce powered
Mk V	Rolls-Royce powered, two built.
Mk VI	Bristol Hercules powered, 1,000+ built
Mk VIC	Coastal Command
Mk VIF	Night fighter, AI Mark VIII radar
Mk VI (ITF)	Interim torpedo fighter
Mk VII	Proposed Australian version, Hercules 26 engines, none built
Mk VIII	Proposed Australian version, Hercules XVII engines, none built
Mk IX	Proposed Australian version, Hercules XVII engines, none built

The Hercules: The Other Engine That Helped Win the War

Mk	Description
TF Mk X	Torpedo fighter, the 'Torbeau', Hercules XVII; 2,231 built
Mk XIC	Coastal Command, no torpedo
Mk XII	Proposed long-range version, none built.
Mk 21	Australian modified version
TT Mk 10	Target tug version

Crew: (Two) Pilot, navigator/radar operator
Max takeoff weight: 11,521 kg, (25,400 lb)
Fuel capacity: 2,500 L (550 imp gal) normal; 3,100 L (682 imp gal) with auxiliary tanks
Powerplant: Two Bristol Hercules XVII or XVIII; 1,200 kw (1,600 hp)
Propellers: 3-bladed constant speed

Performance
- Maximum speed: 510 kmh (320 mph) at 3,000 m (10,000 ft)
- Cruise speed: unknown
- Range: 2,820 km (1,750 miles)
- Service ceiling: 5,800 m (19,000 ft)

Armament
Guns: Four 20 mm (.787 in) Hispano Mark II cannon in nose, six 7.7 mm (.303 in) Browning machine guns in wings, four starboard, two port, one manually operated 7.7 mm (.303 in) Browning machine gun for observer.
Rockets: 8 RP-3 27 kg (60 lb)
Bombs: Two 110 kg (250 lb) or One British 45 cm (18 in) or Mark 13 torpedo

Bristol 170 Freighter

First Flight: 2 December 1945
Built by: Bristol Aeroplane Company, Production: 214
Introduction: 1946

The Hercules Powered Aircraft

As early as 1937, the Bristol Aeroplane Company (BAC), was considering the design of a very large bomber and simultaneously a very large transatlantic airliner. The Brabazon Report in 1943 called for five designs, one of which, the Type I, was the large transatlantic airliner. The BAC final concept design was published in November 1944 of a '100 ton' aeroplane propelled by eight paired Bristol Centaurus eighteen-cylinder engines driving contra-rotating propellers. A daunting company challenge, but BAC was concerned that all their production future was in one design that may not come to fruition, hence the Bristol Type 170.

BAC wanted a suitable design for post-war quick production, both at home and in Australia. Initially, a passenger and freight-carrying derivative of the Bristol Buckingham was considered. However, the final decision was to construct something simpler to build, cheaper to operate, developed from the high-wing, all-metal, monocoque Bristol Bombay. The Bombay was very successful in North Africa, both operationally and economically.

Initial Design: By 1944, the new design had evolved. The Type 170, using the standard Bristol design sequential numbering system, was a rugged, short-range, general duty high-wing transport aircraft. It was to have a low initial cost, be inexpensive to maintain, and not require any special tools. The structure was based on the Bristol Bombay and had a 29.9 m (98 ft) wing span of the same section and taper ratio. However, the Type 170 had swept-back leading edge, straight trailing edge, and simplified two-spar construction; the Bombay had seven spars. The wide track fixed landing gear was retained. A single fin and rudder with trim-tabs completed the empennage. It featured a Sperry autopilot and could be trimmed to fly 'hands-off'. The proposed flat-sided fuselage had horizontally hinged clamshell doors in the nose to accept bulkier freight. These replaced the side doors of the Bombay. The proposed engines were a new version of the Bristol Perseus that had nine Centaurus type cylinders developing 858 kw (1,150 hp) for takeoff.

The Hercules: The Other Engine That Helped Win the War

Development: The Air Ministry recognised the potential of the Type 170 design as a military freighter. It was urgently needed in the Burma Campaign to carry vehicles and supplies into the jungle bases. The one provision that must be met was the requirement to carry a 3.048-kg (3-ton) army truck or equivalent load.

The resulting unobstructed fuselage design, with access through side-hinged nose doors, was 9.8 m (32 ft) long and 2.4 m (8 ft) wide. There was a smaller compartment aft with its own door. The total cargo area was 66.8 cu m (2,360 cu ft). Then came a rethink of the development; the Type 170 would not reach Burma before the end of the war. Cost calculations for the civil market indicated that the aircraft would benefit from more weight and more powerful engines. The war-proven Bristol Hercules engines were readily available and powerful enough to cope with the new weight of 15,876 kg (35,000 lb).

The final decision was to use the Hercules 630 (military 130) with single-speed M ratio blowers giving 1,245 kw (1,670 hp) for takeoff and a range of 1,770 km (1,100 miles) at an economical cruising speed of 241 kmh (150 mph). The landing gear was left at fixed type as there was no advantage in using retractable landing gear for such a short range. The fixed gear was cheaper to make and maintain and no hydraulic system was needed. This allowed the brakes and split-wing flaps to be operated pneumatically.

The post-war civilian aviation industry was demanding flexibility for underdeveloped areas with passenger and freighter versions of the same aircraft. Production in two versions was decided upon, the Freighter with nose doors and an extra strong floor and the Wayfarer with a fixed nose and seating for thirty-six passengers. Both types would have windows, escape hatches, and cabin heating equipment. There was intense competition from war surplus aircraft, which required the Ministry of Supply (MOS) to assist manufacturers of British aircraft to be competitive. Initially, two prototypes were purchased by MOS, two were built as Company demonstrators, and a production order of twenty-five aircraft was authorised.

Armstrong Whitworth Albemarle. (Department of National Defence, Canada (Patrick Martin))

Avro Lancaster B.II. (Imperial War Museum)

Avro York C.II. (Bristol Aeroplane Co.)

Bristol Beaufighter. (Air Force Museum of New Zealand)

Bristol 170 Mk31M Freighter, Stansted 2 June 1996. (Keith Burton)

Right: Bristol 170 Superfreighter, Channel Air Bridge, Southend 1962.

Below: Breguet 890 Mercure. (Public Domain, Avions Legendaires (Dave Beale))

Bottom: CASA C-207 XT.7 Torrejon, Spain 07 June 1973. (John Mounce)

Above: Fokker T.IX. (Public Domain, Pinterest, (Dave Beale))

Left: Folland Fo.108. (Public Domain, Aviation Rapture, (Dave Beale))

Below left: Handley Page Halifax B Mk III. (Air Force Museum of New Zealand)

Handley Page Hastings, Wigram. (Air Force Museum of New Zealand)

Handley Page Hermes, Falcon Airways. (Jerry Hughes)

Nord Noratlas 08 June 2017. (John Mounce)

Northrop 8A. (Public Domain, Duggy009, (Dave Beale))

Northrop Gamma 2L. (Public Domain, (Dave Beale))

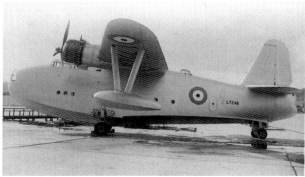

Saro Lerwick. (Seawings)

Short S.26. (Seawings)

Short Seaford. (Imperial War Museum)

Above right: Short Solent. (Air Force Museum of New Zealand)

Right: Short Stirling, RAF No. 1651 Heavy Conversion Unit, RAF Waterbeach. (Air Force Museum of New Zealand)

Below: Vickers Valetta C1, Gibraltar 09 May 1968. (John Mounce)

Vickers Varsity T1, RAFTC Strubby, 16 October 1970. (John Mounce)

Vickers VC.1 Viking 3B, Air Ferry, Liverpool 1964. (John Mounce)

Vickers Wellesley Mk1. (Imperial War Museum)

Vickers Wellington. (Imperial War Museum)

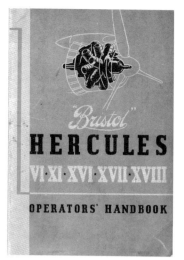

Above left: Bristol Hercules Operators Handbook, August 1945. (Gordon Wilson)

Above right: Handley Page Hastings, Singapore. (Air Force Museum of New Zealand)

Right: Short Empire Flying Boat, 'Cambria'. (Richard Beale)

Below right: Vickers Wellington, Brooklands 15 September 2016. (John Mounce)

Opposite top: Vickers Varsity, RAE, RAF Museum Cosford 21 August 2022. (Bill Powderly)

Opposite middle: Bristol 170 Mk31M, Iraqi Air Force, Shannon Airport, Ireland 26 Dec 1969. (Malcolm Nason)

Opposite bottom: Handley Page Hastings C1A, RAF, Imperial War Museum, Duxford July 2018. (Bill Powderly)

Above: Vickers Viking 1A, BEA Cosford 17 March 1990. (John Mounce)

Above right: Vickers Viking crash, Largs, Ayrshire, Scotland 21 April 1948. (David McNeur)

Right: Short Solent Flying Boat, Museum of Transport and Technology, Auckland May 2019. (John Mounce)

Above: Handley Page Halton. (Board of Aircraft Accidents Archives)

Below: The Bomber Command Museum of Canada, Nanton, Alberta Canada. (Doug Bowman)

Right: Bristol 170 Mk 31 Hawkair, Reynolds-Alberta Museum, Alberta, Canada June 2005. (Malcolm Nason)

Below: Hercules spare parts storage, Hawkair, Terrace, British Columbia Canada. (Karl Kjarsgaard).

Bottom: Karl Kjarsgaard, President, Halifax 57 Rescue (Canada) with Hercules 734 engine.

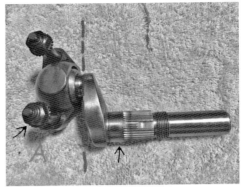

Above left: Hercules cylinder sleeve. (Karl Kjarsgaard)

Above right: Hercules sleeve crank ball joint. (Karl Kjarsgaard)

Left: Hercules sleeve-valve, piston, and cylinder. (Karl Kjarsgaard)

Below left: Hercules sleeve-valve gearing mechanism. (Karl Kjarsgaard)

Right: Single sleeve-valve cylinder operation exhibit. (Karl Kjarsgaard)

Below: Hercules 734 engine cutaway, British Columbian Aviation Museum, Sidney. (Colm Egan)

Above: Hercules 734 engine, airshow display trailer. (Karl Kjarsgaard)

Below: Hercules 734 engine, ground running action. (Bomber Command Museum of Canada)

The Hercules Powered Aircraft

The first prototype was flown at Filton, Bristol, on 2 December 1945 and required some modification to the tailplane to allow trimmed flight over the wide range of centre of gravity positions. It had its nose doors fixed closed and featured round windows instead of the future production rectangular windows. Nine months later, the aircraft was painted in RAF markings and delivered to Boscombe Down for further evaluation.

Meanwhile, the second prototype was constructed as a thirty-two seat Wayfarer and flew on 30 April 1946. Six weeks later it received the first unrestricted Certificate of Airworthiness given to a new post-war aircraft. By October, flying with Channel Airways, it had flown 460 hours carrying 10,000 passengers from Croydon, South London, to Jersey, Channel Islands.

The third prototype, a freighter, left Filton in August 1946 on a demonstration tour of North and South America. Once again, the Hercules engine was crossing the Atlantic, but this time in a fixed landing gear aircraft. From Canada the tour proceeded through America, the Caribbean, Mexico and South America before returning to Toronto, Canada, six months later. Further work in Canada's North and Venezuela was followed by car ferry work and the Berlin Airlift. The Hercules engine had certainly proved itself in the extreme weather conditions. The fourth prototype, a Wayfarer, ended up in the RAF as a flying laboratory for nose radomes. The demonstrated success of the aircraft garnered orders and due to individual requirements a system of Mark numbers was established for future aircraft.

Once again, BAC sent demonstrators to Europe and North Africa and another one to South America. Unfortunately, the South American demonstrator crashed but it did not affect orders from Brazil, two, and Argentina, fifteen. Other customers were REAL, Sao Paulo, in Brazil, Dalmia Jain Airways in Delhi, India, Bharat Airways, Skytravel, all for Mk IIAs and two Mk IICs for Indian National Airways. The remainder of the first production batch were for BEA.

The Bristol 170 aircraft showed its versatility by being used for photographic survey, transporting racehorses, and transporting

The Hercules: The Other Engine That Helped Win the War

refugees (117 with all seats removed), the latter following the partition of India and Pakistan, and landing at Shell Oil Company's high-altitude oilfield in Ecuador. In March 1947, a demonstrator Mk IA went on a sales tour of Australia and New Zealand. The increased span of the Mk XI could not be fully utilised until December 1947 when the more powerful Hercules 672 became available. These new engines and an increase in weight to 18,144 kg (40,000 lb) became the Mk 21.

Two versions were offered, the unfurnished Mk 21 and the convertible Mk 21E. This 'E' version had cabin heating, sound insulation, sixteen pairs of easily removable seats, and a movable bulkhead. It gave the flexibility of the Mks I, IA, and II. Many of the early customers had their aircraft upgraded to this new version during major overhaul, indicating the success of the aircraft and its Hercules engines.

Two accidents were attributed to structural failure of the fin during single-engine operation. The solution was a dorsal fin to prevent rudder-locking during extreme yaw conditions. This new version, in addition to the Hercules 734 engines and an all-up weight of 19,958 kg (44,000 lb), became the Mk 31. The flight development programme was conducted at Boscombe Down, Winnipeg, Canada, and Singapore. The largest single contract for Freighters, Mk 31M, was from the Pakistan Air Force, thirty-eight, for general duties in mountainous terrain.

Perhaps the best-known variant was the Mk 32. It was initiated on a request by Silver City Airways for more car-carrying capability. It was called the 'Superfreighter' and was lengthened by 1.5 m (5 ft) and had an enlarged fin and tailplane. This version could accommodate three medium cars and twenty-three passengers. The first aircraft was delivered in March 1953. Silver City further converted a Mk 32 to a 'Super Wayfarer' with seats for sixty passengers for its London-Paris air-coach service in June 1958. By 1961, the Bristol Hercules engines had hauled around 600,000 cars and 1.5 million passengers over the English Channel to mainland Europe.

By 1963 some of the first aircraft had 26,000 landings and were due for wing spar lower boom replacements. In fact, this procedure could be done more than once, giving the aircraft a life of 60,000 landings. Of the 214 aircraft built, 112 were still in service in December 1963 and had amassed over a million flying hours in world-wide operations and in all extremes of climate. Definitely a winning combination of the Bristol Hercules engine and airframe.

Operators Military: Argentina, Australia, Burma, Canada, Iraq, New Zealand, Pakistan, United Kingdom.

Operators Civil: The Bristol 170 was widely used in many countries, operators in brackets. Argentina, Australia (7), Belgium (2), Brazil, Canada (12), Ecuador, France (10), Germany (2), India (3), Ireland (2), Italy, Laos, Lebanon. New Zealand (2), Nigeria, Rhodesia, Saudi Arabia, Spain (2), South Africa, Sweden, United Kingdom (22), Vietnam.

Variants

Mk	Aircraft Description
I	Freighter, nose doors, heavy floor, crew toilet
IA	As Mk I but combi, 16 seats, one toilet
IB	As Mk I with British European Airways (BEA) requirements
IC	As Mk IA with BEA requirements
ID	As Mk IA with British South American Airways requirements
II	Wayfarer, fixed nose, no seats, crew toilet.
IIA	Wayfarer, 32 seats, pantry, one toilet
IIB	As Mk IIA with two toilets and BEA requirements
IIC	Wayfarer, 20 seats forward, luggage hold, toilet aft
XI	Freighter: Span 33 m (108 ft), larger fuel tanks, weight 17,690 kg (39,000 lb), Hercules 632 engines
XIA	Mixed-traffic version
21	As Mk XI, Hercules 672 engines
21E	As Mk I, IA, IIA, cabin heating, sound insulation, 32 removable seats, with adjustable bulkhead.
21P	Specially for Pakistan Air Force, windows in nose section

The Hercules: The Other Engine That Helped Win the War

Mk	Aircraft Description
22A	Wayfarer, fixed nose, intended version of Mk 21, none built.
31	Freighter with Hercules 734 engines.
31C	Special version for A&AEE, Boscombe Down.
31E	Freighter as per 21E.
31M	Freighter, military version.
32	Car Ferry, extended nose and modified tail assembly.

Mk I, IA, II, IIB, IIC, XI, XIA
Crew: (Three)
Max takeoff weight:
I, II   16,556 kg (36,500 lb)
IA, IIA, IIC   16,783 kg (37,000 lb)
XI, XIA   17,690 kg (39,000 lb)
Powerplant: Two Bristol Hercules 632 1,249 kw (1,675 hp)
Propellers: 3-bladed de Havilland Hydromatic

Performance
- Maximum speed: 386 kmh (240 mph), except I & XIA 314 kmh (195 mph)
- Cruise speed: 262 kmh (163 mph)
- Range: 966 km (600 miles) except XI, XIA 1,448 km (900 miles)
- Service ceiling: 6,706 m (22,000 ft) except XI, XIA 5,791 m (19,000 ft)

Mk 21, 21E, 21P
Crew: (Three)
Max takeoff weight: 18,144 kg (40,000 lb)
Powerplant: Two Bristol Hercules 672 1,260 kw (1,690 hp)
Propellers: 3-bladed de Havilland Hydromatic

Performance
- Maximum speed: 362 kmh (225 mph)
- Cruise speed: 262 kmh (163 mph)

- Range: 1,448 km (900 mi)
- Service ceiling: 6,401m (21,000 ft)

Mk 31, 31C, 31E, 31M, 32
Crew: (Three)
Max takeoff weight: 19,958 kg (44,000 lb)
Powerplant: Two Bristol Hercules 734 1,477 kw (1,980 hp)
Propellers: 3-bladed de Havilland Hydromatic

Performance
- Maximum speed: 362 kmh (225 mph)
- Cruise speed: 262 kmh (163 mph)
- Range: 1,320 km (820 mi)
- Service ceiling: 7,468 m (24,500 ft)

Fairey Battle

First Flight: 10 March 1936
Built by: Fairey Aviation Company, Production: 2,201
Introduction: June 1937

The Fairey Battle, a light bomber, was designed to replace the aging biplane Hawker Hart and Hawker Hind biplanes in the mid 1930s. It was a much heavier aircraft than its predecessors and used the Rolls-Royce Merlin engine. It really was too little, too late for the times, being too slow, having limited range, and insufficient defensive fire power. It was used at the start of the war and during the 'Phoney War' stage had the distinction of scoring the first aerial victory for the RAF. It suffered way above average losses during Advanced Air Striking Force operations in France in 1940 and by the end of the year was withdrawn from front-line service. It was a great disappointment as a bomber, but it redeemed itself somewhat when it was relegated to training units overseas. Its association with the Bristol Hercules was that it was used as an engine test bed for the Hercules II and XI engines.

The Hercules: The Other Engine That Helped Win the War

Development: was in response to Air Ministry specification P27/32 for a day monoplane bomber. Marcel Loebelle, the principal designer favoured the Rolls-Royce Merlin engine due to its smaller frontal area and power profile. The aircraft had clean lines which was added to by a retractable tail wheel. Of the various company submissions, Fairey and Armstrong Whitworth were selected to build a prototype. The Fairey Battle flew in March 1936 and, because of the concern over the looming war, was immediately transferred to RAF Martlesham Heath for service trials and subsequently, without delay, an initial production order was generated for 155 aircraft. Some members of the Air Staff expressed an opinion that it was a rushed decision to fill a need with an unsuitable aircraft due to previous indecision; they were right.

In June 1937, the initial production aircraft had its first flight and by the end of 1937, eighty-five Battles had been built and some had reached RAF squadrons. Further to the pre-war expansion programme, a total of over 2,000 aircraft were on order, a case of quantity over quality. The reality was that the Battle was the 'only show in town' and the momentum of production by Fairey at Stockport, Cheshire, and the Austin Motor Company shadow factory at Longbridge (Birmingham), West Midlands, continued despite performance concerns. A contributing factor was the huge demand for skilled labour and material resources on the country; the Battle was already up and running with personnel and supply lines in place. By September 1940, reality finally won out and all production was cancelled. The Battle did at least play a part by being operational during the 'Phoney War' and then subsequently by filling a vital training role.

Design: the Battle was a single-engine, Rolls-Royce Merlin, low-mounted cantilever monoplane. Production aircraft were progressively powered by various versions of the Merlin engine. The two-section fuselage was a slim, oval configuration. The steel tubular forward section in front of the cockpit supported the weight of the nose-mounted engine; the rear section was a metal monocoque structure built on jigs. It was an innovative construction for the Fairey Company. It was the Company's first low-wing monoplane and first use of a light-alloy stressed skin design. The centre section of the wing was an integral part of the fuselage. The internal structure of the wing had steel spars with the ailerons, elevator, and rudder being metal framed with fabric coverings. The split trailing edge flaps were metal.

The pilot and gunner were seated in tandem in the cockpit with the observer positioned directly below the pilot's seat. The pilot operated the fixed gun in the wing and the gunner operated the rear-facing manually operated Vickers K gun. The Mk VII Course Setting Bomb Sight was used by the bomb aimer through a sliding panel in the floor of the fuselage. The bombs were carried internally within cells in the wings, or externally under bomb bays, or on wing mounted racks. The bombs were mounted on hydraulic jacks and released through trap doors.

The Battle was a robust aircraft with a reputation of being relatively easy to fly for novice pilots although it did have one ergonomic challenge, operating the flaps and landing gear together. The landing gear had an awkward safety catch. The cockpit was roomy with good visibility for the pilot. Unfortunately, it was already obsolete by its introduction. The Battle had improved greatly over its biplane antecedents, but the fighters had improved even more. It lacked the self-sealing fuel tanks that subsequently became a necessity for survival in the air. It was quickly taken out of front-line service before it had the chance to undergo such modifications. It was an aeroplane caught out by the rapidly advancing technology at the start of the war.

Operational history: In June 1937, No. 63 Squadron RAF, RAF Upwood, Cambridgeshire, became the first squadron to be equipped with the Fairey Battle. It was the first aircraft type with the Rolls-Royce Merlin engine to enter service, just months ahead of the Hawker Hurricane. Two years later, seventeen squadrons were operational with the Battle. In September 1939, ten squadrons were deployed with the RAF Advanced Air Striking

The Hercules: The Other Engine That Helped Win the War

Force to France. They were in position to offer retaliatory bombing attacks in case of German bombing attacks on France. The first RAF aerial victory is attributed to the Battle shooting down a Messerschmitt Bf 109 near Aachen, Germany, on 20 September. This victory was the exception. The Battles losses were exorbitant: on the first sortie three out of eight, second sortie ten out of twenty-four, and during an attack on the Albert Canal, seven out of eight. A further operation at Sedan on the river Meuse resulted in thirty-five losses out of sixty-three aircraft; the Battle was immediately mainly switched to night operations. By June 1940, the Battles returned to England with a total loss of 200 aircraft. If it was not for the bravery of the Battle crews, facing overwhelming odds, nothing meaningful would have been accomplished. Two Victoria Crosses were awarded during the Albert Canal action. By October 1940, all bombing operations ceased, and the aircraft reverted to some coastal reconnaissance patrols and mainly training and glider towing roles. The Battle saw service with the South African Air Force in Ethiopia, Italian Somaliland, and Eritrea and very limited service in Greece and Turkey.

Training Role: The Battle found a new important role as a dual control pilot training aircraft, some equipped aircraft as a gunnery/bomb aimer trainer and as target towing aircraft for aerial gunnery training. A total of 802 Battles went to the British Commonwealth Air Training Plan in Canada. These aircraft were gradually replaced by the more advanced Bristol Bolingbroke and North American Harvard during the war. Similarly, some aircraft went to Australia.

The Bristol Aeroplane Company used the Fairey Battle as an engine test bed for their Hercules II and XI engines. Its relatively stable handling characteristics made it ideal for this, although I have always wondered about the wisdom of using a single engine aircraft to test engines! A variety of engines are known to have been tested by various manufacturers: Rolls-Royce Exe, Fairey Prince and Monarch, Napier Dagger and Sabre, and Wright Cyclone R-1820.

The Hercules Powered Aircraft

Variants

Mk	Aircraft Description
Day Bomber	Prototype (K4303)
I	Three seat light bomber, Rolls-Royce Merlin I 770 kw (1,030 hp)
II	Rolls-Royce Merlin II 770 kw (1,030 hp)
III	Rolls-Royce Merlin III
V	Rolls-Royce Merlin V
T	Mk I, II, and V converted to trainers.
IT	Mk I, II, and V converted to trainers, rear turret.
IIT	Mk I prototype conversion, 630 kw (840 hp) Wright Cyclone R-1820-G38
TT	Target tug, last model produced, 226 built.

Operators Military: Royal Australian Air Force (number of aircraft 366), Belgian Air Force (16), Royal Canadian Air Force (739), Indian Air Fore (4), Irish Air Corps (1), Hellenic Air Force (12), Polish Air Force, South African Air Force (340), Turkish Army Air Force (30), RAF (26 Squadrons), Fleet Air Arm (3).

Mk III
Crew: (Three) pilot, observer (bomb aimer), radio operator/gunner.
Max takeoff weight: 4,895 kg (10,792 lb)
Fuel capacity: unknown
Powerplant: One Rolls-Royce Merlin II 770 kw (1,030 hp)
Propellers: 3-bladed

Performance
- Maximum speed: 414 kmh (257 mph) at 4,572 m (15,000 ft)
- Cruise speed: unknown
- Range: 1,600 km (1,000 mi)
- Service ceiling: 7,600 m (25,000 ft)

Armament
Guns: one fixed, forward-firing .303 in (7.7 mm) Browning machine gun in starboard wing, one flexibly mounted .303 in (7.7 mm) Vickers K machine gun in rear cockpit.
Bombs:
Internal 450 kg (1,000 lb)
External 680 kg (1,500 lb)

Folland Fo.108

First Flight: 1940
Built by: Folland, Production: 12
Introduction 1940–1946

The Folland Fo.108 was specifically designed in response to an Air Ministry specification calling for a dedicated aircraft to be an engine test bed for the aero-engine industry. It was a breakthrough for the Folland company as it was their first design to be accepted for production.

Development: It used the Bristol Hercules engine as the ferry engine to power the aeroplane to its next engine assignment with the aero-engine companies.

Design: It was a large low-wing cantilever monoplane with a conventional cantilever tailplane. The landing gear was non-retractable. The purpose-built aeroplane had a glazed cockpit for the pilot and a special cabin, behind and below the pilot, for two observers to monitor and chart engine performance during the test flight.

Operational history: It tested the H-24, liquid-cooled, sleeve-valve Napier Sabre engine; the two-row, 18-cylinder, air-cooled, radial Bristol Centaurus; and the sixty-degree, liquid-cooled, V-12 Rolls-Royce Griffon engines. Five of the production aircraft were lost in crashes. The type earned the nickname 'Frightener' as a result.

Operators Civil: Bristol Aeroplane Company, Napier, Rolls-Royce, and De Havilland.

Fo.108 (Centaurus engine)
Crew: (Three) pilot, two observers
Max takeoff weight: 7,257 kg (16,000 lb)
Fuel capacity: unknown
Powerplant: One Bristol Centaurus
Propellers: 4 bladed constant-speed

Performance:
- Maximum speed: 470 kmh (292 mph) unverified
- Cruise speed: 430 kmh (267 mph) unverified
- Range: unknown
- Service ceiling: unknown

Handley Page Halifax

Hercules XVI, Halifax B Mk III/VI
First Flight: 25 October 1939
Built by: Handley Page, Production: (III) 2,091/(VI) 643 Total 6,176; 2 HP57 Prototypes
Operational: 13 November 1940

Germany and Italy had begun to rearm and some of the weapons were 'tested' in the Spanish Civil War. The Air Ministry in response issued specification B1/35 for a twin-engine replacement for medium bombers, such as the Armstrong Whitworth Whitley and Vickers Wellington. The specification called for a 907 kg (2,000 lb) bombload to be delivered at 314 kmh (195 mph) at 4,572 m (15,000 ft) over 2,414 km (1,500 miles). An interesting restriction as the country approached a possible war was a maximum 31 m (100 ft) wing span so that the aircraft would fit in the standard RAF hangar! It was to be powered by two engines rated at 746 kw (1,000 hp).

The Hercules: The Other Engine That Helped Win the War

Handley Page proposed their HP55 design with a low mounted wing, swept back wing leading edge, and landing gear retracting into the engine nacelle. The government specified as small an aircraft as possible, so the bomb load was split between the fuselage and bomb cells in the two-spar wing. The proposal was rejected. The next specification issued was P13/36 for a 20,412 kg (45,000 lb) all-metal, mid-wing, cantilever monoplane, medium bomber to be powered by Rolls-Royce Vulture engines with the ability to carry a 3,629 kg (8,000 lb) bomb load. Handley Page designer George Volkert realised that an aircraft designed for B1/35 would meet the new requirements. A mock-up of the proposed Handley Page HP56 was built. The specifications were continually changing but work had started on the first prototype until it became apparent that the Rolls-Royce Vulture engines would not be available in sufficient quantities. The Vulture engine has twenty-four cylinders arranged in four banks of six cylinders in an X configuration. Both proposals, HP55 and HP56, were never built.

Development: Handley Page was then asked to investigate various four-engine layouts using the Bristol Taurus, Rolls-Royce Kestrel, Napier Dagger, and Bristol Pegasus engines. Interesting that the Bristol Hercules engine was not considered, as it had been allocated to the Short Stirling bomber. None of these engines was capable of the performance demanded of a four-engined bomber and the Handley Page aerodynamic and performance team concluded that it would take the up-rated Rolls-Royce Merlin engines to meet the requirement. In September 1937, the Air Ministry ordered two prototypes powered by four Rolls-Royce Merlin engines.

The basic design underwent many alterations, such as increased wingspan and fuselage length, increased weight to 18,144 kg (40,000 lb), centre of gravity brought forward, rear spar straightened, bomb bays installed inboard of the inner engines, integral fuel tanks between the inner and outer engines, and to accommodate all these changes, the wing area was increased by

The Hercules Powered Aircraft

twenty-one per cent. This redesign caused production delays as war loomed.

The smaller diameter of the Merlin propellers in comparison the proposed Vulture engines resulted in lower cruise speed and shorter range. The increased weight resulted in larger landing gear wheels, which increased drag. Two other factors also increased drag, the thicker wing section to house the bomb bays and the deep fuselage with its large frontal area.

In January 1938, construction began on the first prototype at the Handley Page Cricklewood factory and was completed just before the start of the war in September 1939. It was designated the HP57 and given the name of Halifax, as per company practice to name heavy bomber aircraft after towns. The prototype had the turret positions faired in, De Havilland three-bladed metal propellers, and integral wing tanks for the first flight. Similar to its predecessor, the Handley Page Hampden, the Halifax had leading edge slots on its outer wing panels to lower stall speed.

The first flight was at RAF Bicester, Oxfordshire, to facilitate assembly of the major components. Temporary fuel tanks were installed in the bomb bay as the integral fuel tanks were filled with water ballast to facilitate full load and overload trials. There was a system to dump the water ballast to prevent an overweight landing in case of an immediate return after takeoff. A late change of brake system from the Lockheed hydraulic system to the Dowty pneumatic system was made due to the slow action of the Lockheed system found during taxiing trials. The brake system was not completely changed for the first flight and relied on a pair of compressed air bottles in the fuselage.

A first flight, landing gear down, was flown after the war had started on 25 October 1939. The aircraft, L7244, performed well with little modification and was then flown to A&AEE (Aeroplane and Armament Experimental Establishment) at Boscombe Down, Wiltshire, for further testing. L7244 was later grounded and used as an instructional airframe after it lost three of its propellers. The second prototype, L7245, flew in August 1940, complete with turrets, at a weight of 22,680 kg (50,000

The Hercules: The Other Engine That Helped Win the War

lb). It had constant speed Rotol propellers with Schwartz wooden blades for comparison purposes. It reversed the test configuration by having fuel in the fuel tanks and the water ballast carried in the bomb bay. It, too, went to A&AEE and narrowly avoided being written off due to a stuck starboard landing gear causing a burst accumulator.

The first of the 100 production aircraft ordered, L9845, flew in August 1940. Although still retaining the leading edge slots, subsequent aircraft were fitted with barrage balloon cable cutters. To speed production, orders were also allocated to the English Electric Company at Preston, Lancashire, Fairey Aviation Co. at Errwood Park, Derbyshire, Rootes Securities at Speke, Liverpool, and the London Aircraft Production Group at Leavesden, Hertfordshire. Some of these companies had no previous experience of aircraft production. The manufacturing arrangement worked because the aircraft structure consisted of fifteen main split sub-assemblies, which simplified production. 1,967 Merlin-powered aircraft were built. About one in a hundred aircraft were flight tested by company pilots at Radlett to ensure standards, but it was wartime.

The development of the Halifax would continue with emphasis on the armament and power plant components. The armament varied between combinations of .50 calibre guns, .303 machine guns, and beam amidships guns on both sides. A dorsal gun was never implemented. The powerplants would favour both the Rolls-Royce Merlin and the Bristol Hercules. The Halifax was proposed as a replacement for all Short Stirling and Avro Manchester aircraft. However, Avro had converted the Manchester into a very successful four-engined Rolls-Royce Merlin powered aircraft called the Avro Lancaster.

The Halifax leading edge slats were wired shut and would eventually be deleted on the B Mk I Series III, as Handley Page strived for performance. There were other development difficulties. The second series B Mk II, with its higher wing loading, experienced stability problems. There were also Messier landing gear problems. The self-centring retractable tail wheel

The Hercules Powered Aircraft

was unreliable and was locked down for night landings where it could not be checked for visual alignment. The main landing gear was slow to retract, which reduced takeoff performance and it had the added disadvantage of being time-consuming to manufacture.

There was a proposal to use a Dowty landing gear system, the same as on the Avro Lancaster, however, there was a compatibility drawback. The Messier system used a castor oil system with natural rubber seals and the Dowty system used mineral oil with synthetic rubber seals. The compromise was a Messier system with mineral oil and neoprene seals. To alleviate the landing gear shortage, the solution was to install Dowty landing gear in a new production number, the HP63 Halifax B Mk V. However, instead of using the original developed forgings it was decided, to save time, to use castings. These castings developed brittle fractures with the result that the maximum landing weight had to be reduced to 18,144 kg (40,000 lb). This aircraft was used by six RCAF squadrons, one RAF squadron, and two Free French squadrons. The B Mk V ended production in January 1944 after 904 aircraft had been built.

An unusual performance enhancement on the HP59 Halifax B Mk II and HP63 B Mk V aircraft was the removal of the Boulton Paul Type C nose turret, replaced with 'Z' fairing, and removal of the Type C1 mid upper turret. This weight saving resulted in a speed increase of 26 kmh (16 mph). The idea had been put forth that the turrets were of use in daytime only, certainly a debatable point! The Special Operations Executive needed the speed for their supply and agent dropping operations in support of the Polish resistance.

This was not the only enhancement. A streamlined nose, low drag engine nacelles from the Bristol Beaufighter, and Vickers K nose gun were added. Coastal Command replaced this gun with a belt-fed .50 calibre gun, ventral FN 64 turret, and additional 3,137 L (690 imp gal) fuel tanks installed in the bomb bay. Special radar and meteorological instruments were added, depending on role.

A series of modifications were made to the metal bomb bay doors, changing them to faired wooden doors on steel frames that now allowed the 1,814 kg (4,000 lb) bomb to be carried. A major modification to these two aircraft was to enlarge the vertical tail to correct rudder stalling in the event of two engines failing on the same side. It took many trial and error versions to find the solution. In an effort to return the Halifax to daylight operations, a high-altitude version was considered using the Merlin 60 series or the Bristol Hercules VIII engine. Unfortunately, the Hercules was plagued by supercharging surging problems and was dropped.

In September 1941, the government proposed a Bristol Hercules VI or XVI powered version, the HP61 Halifax B Mk III, with a weight of 29,030 kg (64,000 lb) with an estimated speed of 494 kmh (307 mph) at 6,401 m (21,000 ft) altitude. The Bristol Hercules 100 engines would be installed at a later date. Handley Page was requested to incorporate several modifications such as provision for parachute drops, protective engine cowlings, multiple flare chutes, and arrester gear for landing. These were all deferred because it would interrupt the production schedule too much, but a Boulton Paul Type A mid upper turret and FN 64 ventral turret for daylight operations were installed, as were engine exhaust flame dampers, engine rear armour, D/F loop, provision for glider towing, and increased electricity generation.

The Hercules-powered HP61 B Mk III was the most numerous variant with 2,127 aircraft being built by all five main assembly groups. What were its origins? In January 1942, a B Mk II Series I (Special) was selected for installation of the Hercules VI engine. The prototype first flew in October with a few developmental changes such as a retracting tailwheel, Beaufighter type Hercules engine cowlings, extended wing tips, and the ventral gun replaced by H2S radar. These changes were all completed by August 1943 and the first production aircraft were issued to 433(B) Squadron RAF and 466(B) Squadron RAAF. Forty-one operational units would use this variant.

Shortly thereafter, the 'power egg' Hercules 100 became available using 100 octane fuel with RAE-Hobson injection carburettors. It did require a high-pressure fuel supply from seven wing fuel cells of (2,190 imp gal) and additional (690 imp gal) from the bomb bay. Larger oil coolers, a cabin heating system, tropicalisation, and further retrospective modifications meant that this variant was renamed as the HP61 B Mk VI; 457 were built. When demand outstripped supply the Hercules XVI engines were installed in the new variant, HP61 B. Mk VII; 423 aircraft. These variants were available 1943/1944 and were the last variants produced. Those 'tropicalized' were for the Japanese campaign.

Improvements to the Halifax, like other aircraft, were always a tricky balancing act between getting the plane in the air fast versus taking the time to evaluate performance improvements properly. It was structurally robust and could survive extensive flak and fighter damage. Perhaps slightly underpowered to begin with using the Rolls-Royce Merlin engines, the Bristol Hercules engines greatly improved the Halifax performance.

Design: The slab-sided fuselage featured a two-gun nose turret and four-gun tail turret. The deep bomb bay extended the full length of the fuselage centre section and bomb cells were built into the inner sections of the wings between the two main spars. The mid-wing had dihedral on its outer panels, automatic slots on the outer leading edges, and squared wing tips. The tailplane had twin rudders and the original vertical stabilisers were swept back which was later found to cause loss of control. All flight controls were fabric covered.

Operational history: The first Halifax squadron, No. 35 Squadron RAF, had been formed at Boscombe Down on 05 November 1940. The squadron began training on the first prototype, which had been fitted with temporary dual controls. It moved to its operational base at Linton-on-Ouse, Yorkshire, in December, and by March 1941 six crews had qualified. In April, C Flight was expanded to become No. 76 Squadron RAF based at Middleton St George, Durham. In

1942 the Halifax formed the foundation of the Pathfinder Force, later to become No. 8 Group. By 1943, No. 4 Group, Bomber Command, had been entirely equipped with the Halifax.

The HP57 Halifax B Mk I Series II, with its amidship Vickers guns, was thought to be able to defend itself in day operations. However, in the summer of 1941 there was a thirty per cent loss rate on two daylight raids, so daylight operations were suspended for six months.

The Bristol Hercules and the Handley Page Halifax contributed significantly to many wartime operations and eventual victory. Handley Page and Fedden would certainly be proud their success. Halifaxes flew a total of 82,773 operations dropping 203,397 tonnes (224,207 tons) of bombs for a loss of 1,833 aircraft.

Variants:
HP57
B Mk I Series I: 50
B Mk I Series II: 25 (60,000 lb), twin Vickers .303 K guns amidship, both sides
B Mk I Series III: From the seventy-fifth aircraft on, they had large oil coolers to accommodate the Rolls-Royce Merlin XX engines, more fuel, additional generator, and new radios.

HP58
Mk II was a heavily armed day bomber with dorsal and ventral twin cannon amidships and a tapered tail. Developing turrets caused delays, redesignated HP59 B Mk II Series I.

HP59
B Mk II Series I: Fitted with TR 1335 navigation aids.
B Mk II Series I (Special) SOE: Nose armament and dorsal turret removed.
B Mk II Series I (Special): As above with varied combinations of armament.
B Mk II Series IA: Glazed nose, new 'D' fin and rudder, some H2S radar.

B Mk II Series I, Freighter: Simple mods to allow carriage of engines and Spitfire fuselages.
B Mk II Series II: Single aircraft, three-bladed ROTOL propeller, Merlin 22 engines. Rejected.
A Mk II: Perhaps some converted for airborne use?
GR Mk II: Coastal Command variant.
GR Mk II Series I: Coastal Command with some ASV Mk 3 radar and 12.7 mm (.50 in) in nose.
GR Mk II Series IA: Definitive Coastal Command, extra fuel, ASV Mk 3 radar/ventral gun
Met Mk II: Unconfirmed.

HP61

B Mk III: Main variant, increased wingspan 31.75 m (104 ft 2 in), dorsal/turret four guns.
A Mk III: Horsa and Hamilcar glider tug and paratroop transport.
C Mk III: Military transport version, Nine stretchers or eight passengers.
GR Mk III/IIIA: Coastal Command; standard bomber/specific maritime equipment.
Met Mk III: Five Met squadrons during 1945. Met Officer plus specialised equipment.

HP63

B Mk V: Merlin XX engines, square empennage, and wingtips.
B Mk V Series I (Special)
A Mk V: Glider tug and paratroop transport.
GR Mk V: Coastal Command maritime reconnaissance.
B MK VI: Hercules XVI, H2S radar, no dorsal turret, round wing tips, 643 built.
C Mk VI: Military transport version.
GR Mk VI: Coastal Command maritime reconnaissance.
B Mk VII: Hercules XVI, round wing tips.
A Mk VII: Glider tug and paratroop transport.
C Mk VII: Military transport version.

HP 70
C Mk VIII: Cargo and passenger version.

HP 71
A Mk IX: Glider tug and paratroop transport

Operators Military (Squadrons): RAF (42), RAAF (3), RCAF (15), Royal Egyptian Air Force, Free French Air Force (2), Pakistan Air Force (1), Polish Air Force in exile in Britain (1), one flight.

Operators Civil: Australia, France, Norway, Pakistan, South Africa, Switzerland, Britain: Air Freight, Airtech, Bond Air Services, British American Air Services, BOAC, Chartair, CL Air Surveys, Eagle Aviation, Lancashire Aircraft Corporation, London Aero and Motor Services, Payloads, Skyflight, Union Air Services, VIP Services, Westminster Airways, and World Air Freight.

B Mk III
Crew: (Seven) pilot. co-pilot/flight engineer, navigator, bomb aimer, radio operator/gunner, two gunners.
Max takeoff weight: 29,484 kg (65,000 lb)
Fuel capacity: 12,220 L (2,688 imp gal)
Powerplant: Four Bristol Hercules XVI 1,204 kw (1,615 hp)
Propellers: 3-bladed De Havilland Hydromatic metal

Performance
- Maximum speed: 454 kmh (282 mph) at 4,115 m (13,500 ft)
- Cruise speed: 367 kmh (228 mph) at 6,096 m (20,000 ft)
- Range: 2,993 km (1,860 mi)
- Service ceiling: 7,315 m (24,000 ft)

Armament
Guns: Single Vickers K .303 machine gun in nose, four Browning .303 machine guns in Mid-Upper dorsal turret, and four browning .303 in tail turret

Bombs: 5,897 kg (13,000 lb) in various configurations, example: One 227 kg (500 lb) in each of six wing bays, two 907 kg (2,000 lb) and six 453 kg (1,000 lb) in fuselage

Handley Page Hp 70 Halton
First Flight: Post War civil conversion
Built by: Handley Page, Production: 1 Maharajah Gaekwar of Baroda, 12+ conversions
Introduction: 1945
Operators Civil: India (1), France (1), South Africa (1), United Kingdom (12).
Specifications: Unavailable.

Handley Page Halifax Civil Transport (HCT)
Introduction: Last years of war
Built by: Handley Page, Production: 99
The Handley Page Halifax Civil Transport was brought into service by converting the Halifax bomber for civilian use. It had seating for eleven passengers with some seats convertible to sleeping berths. Each passenger seat had a fuselage window and there were toilet facilities in the aft fuselage. The cabin was lined and upholstered throughout. The aircraft also had the facility to carry cargo in a large, streamlined pannier fitted into the former bomb bay. It could carry 3,632 kg (8,000 lb) and had loading hatches fore and aft. It could be lowered and raised by means of a winch to facilitate loading or could be quickly switched with another pannier for a quick turnaround. When there was the need for extreme range flights, the pannier could be replaced by additional fuel tanks with the penalty of reduced payload.

Halifax Civil Transport
Crew: Unknown
Max takeoff weight: 30, 870 kg (68,000 lb)
Fuel capacity: 12,220 L (2,688 imp gal)
Powerplant: Four Bristol Hercules 100 1,249 kw (1,675 hp)
Propellers: 3-bladed de Havilland three-bladed, constant speed

Performance

- Maximum speed: 512 kmh 320 (mph)
- Cruise speed: 416 kmh (260 mph) max weak mixture, 3,050 m (10,000 ft)
- Range: normal fuel tanks and load of 4,880 kg (10,750 lb) 3,872 km (2,420 mi)
- Service ceiling: unknown

Handley Page HP67 Hastings

First Flight: 07 May 1946
Built by: Handley Page, Production: 151
Operational: September 1948–1977

The Handley Page Hastings was a troop-carrier and freighter aircraft designed specifically for the RAF. When it was introduced into service in 1948 it was the largest transport aircraft in the RAF inventory. The government specification had been initiated for a large, four-engined military aircraft in 1944. Up until that time Handley Page had been working on a priority civil airliner for the anticipated travel market. Unforeseen circumstances changed the company direction and the prototype flew in May 1946. The tail required some modifications to improve flight characteristics. The Hastings was just in time to assist with the Berlin Airlift, delivering 49,900 tonnes (55,000 tons) to the beleaguered city. It would eventually take over from Transport Command's Avro York and become the standard long-range transport aircraft. It contributed immensely to the confrontations in the Suez and Indonesian regions. It also saw duty in weather forecasting, training, and as VIP transport. The Hastings would continue in service use until 1977 until replaced with the new turboprop aircraft.

Design: The large four-engined transport aircraft had hydraulically operated landing gear and a cavernous interior, which was readily adaptable to carry various types of loads. The 85 cu m (3,000 cu ft)

The Hercules Powered Aircraft

interior had a special Plymax floor that could be completed with various fittings for securing personnel and cargo loads. This would include paratroopers, stretchers, sitting casualties, or freight. A large freight door was on the port side, which incorporated a paratrooper door. There was also a paratrooper door on the starboard side. Vehicles could be loaded/unloaded by a ground deployed ramp. The circular, rolled alloy fuselage was constructed in three parts and measured 3.4 m (11 ft) in diameter for most of its length. To facilitate outsize loads, strong fixture points were installed on the underside of the fuselage. The low-mounted two-spar cantilevered wing was smoothly faired to the fuselage. The fuel tanks were housed inboard of the inner engines and had an emergency fuel jettisoning system. The four Bristol engines were housed in interchangeable 'power egg' units within the dihedral, all-metal wing and featured an automated air cleaner and fire suppression system.

Development: The Hastings emerged from the latter stages of the Second World War as Handley Page was trying to remain competitive as an aircraft manufacturer. The circular fuselage was capable of sustaining a 38 kPa (5.5 psi) pressure differential for pressurisation. As mentioned above, it took precedence over the Hermes civilian transport as it had crashed on its first flight in December 1945. Five months later the Hastings took to the air with some lateral instability problems and, more importantly, a lack of impending stall characteristics. The stall problem was cured by a new tailplane design. Six aircraft of the initial order were converted to meteorological reconnaissance. Eight aircraft were later converted to T5s to replace the Avro Lincoln Bomber Command trainer. Other major changes were additional fuel tanks within the outer wing and further fuel to make the C4 VIP transport.

Operational history: The Hastings had been rushed into service to supplement, and then replace, the Halifax aircraft with No. 47 RAF Squadron on the Berlin Airlift. Amazingly, the Hastings was used principally to transport coal to the city. About

The Hercules: The Other Engine That Helped Win the War

one hundred aircraft were deployed on the Transport Command long-range routes and overlapped the introduction of the Bristol Britannia until 1959. No. 24 RAF Squadron was assigned four of the VIP-configured aircraft. In 1956, the Hastings dropped paratroopers on the Egyptian El Gamil airfield during the Suez Crisis. It saw service in Indonesia and finally left Transport Command in 1968 to be replaced by turboprop aircraft. The weather reconnaissance Hastings were based in Aldergrove, Northern Ireland, until replaced by satellites in 1964. Deployed during the Cod War with Iceland, the Hastings was withdrawn from service in 1977. The Hastings also saw overseas service with two squadrons of the Royal New Zealand Air Force.

Variants

Model	Description
HP67 Hastings	Two prototypes
HP67 Hastings C1	Production aircraft, Bristol Hercules 101, converted to C1A and T5 (number built 94)
HP67 Hastings C1A	C1 rebuilt to C2 standard.
HP67 Hastings Met1	Coastal Command weather reconnaissance (6)
HP67 Hastings C2	Bristol Hercules 106, enlarged tailplane, increased fuel capacity. (43)
HP67 Hastings T5	RAF Bomber Command, Navigation Bombing System training. (8)
HP94 Hastings C4	VIP version. (4)
HP95 Hastings C3	Bristol Hercules 737, Royal New Zealand Air Force. (4)

Operators: New Zealand, RNZAF (2),* United Kingdom, RAF Squadrons (18), Operational Conversion Unit (3), Communication Squadron (2), Bombing School (2), Central Signals Establishment, Aeroplane and Armament Experimental Establishment, Royal Aircraft Establishment, Meteorological Research Flight.

---

* Number of units/schools/squadrons

Hastings C2

Crew: (Five) Pilot, co-pilot, radio-operator, navigator, flight-engineer.
Max takeoff weight: 36,287 kg (80,000 lb)
Fuel capacity: 14,420 L (3,172 imp gal)
Powerplant: Four Bristol Hercules 106 1,249 kw (1,675 hp)
Propellers: 4-bladed de Havilland Hydromatic

Performance

- Maximum speed: 560 kmh (348mph) at 6,800 m (22,200 ft)
- Cruise speed: 468 kmh (291 mph)
- Range: 6,840 km (4,250 miles) maximum fuel, 3,400 kg (7,400 lb) payload
- Service ceiling: 8,100 m (26,500 ft)

Handley Page HP81 Hermes

First Flight: 02 December 1945
Built by: Handley Page, Production: 29
Introduction: 06 August 1950.

The Handley Page Hermes was a civilian airliner designed during the final years of the war. It was designed in parallel with the military Hastings. It was intended to introduce the Hermes to airline service in advance of the Hastings but was delayed by the fatal loss of the prototype during its maiden flight. Immediate measures were taken to cure the longitudinal instability problem and add to its carrying capability. In 1947, BOAC ordered twenty-five HP81 Hermes IV aircraft. It entered airline service in 1950 after a delay to reduce the empty weight.

Design/ Development: Two years before the end of the Second World War, the Air Staff were looking for a replacement for the Halifax transport/freighter and also for an interim civil airliner for the post-war years, unusual optimism for 1943. The

The Hercules: The Other Engine That Helped Win the War

specification for the civil aircraft called for a pressurised aircraft carrying thirty-four first class or fifty tourist passengers. The Hermes would fly first, with the Hastings, which shared many of the components with the Hermes, second. One of the obvious differences between the aircraft was the Hermes's tricycle gear and the Hastings's tailwheel configuration. It is believed the crash of the first prototype was due to elevator overbalance. The delay in production allowed the second prototype to have its fuselage extended. In September 1948, it received its Certificate of Airworthiness. A Hermes V proposal was to equip it with four Bristol Theseus turboprop engines to take over from the Hercules engines and make it competitive with such aircraft as the Bristol Britannia. Two prototypes were flown and used for research purposes, the first later destroyed in a landing accident.

Airline history: The Hermes had a chequered entry into airline service with BOAC, its initial customer, who said that its weight compromised its performance. This weight gain was partially from the modifications to improve stability and the use of heavier Hastings components. Design alterations were made, including lightening the floor structure. Finally, in August 1950, the Hermes took over the England to African routes from the Avro York. It had a short life as it was replaced by the Canadair Argonaut with Roll-Royce Merlin engines in 1952. It momentarily came back into service during the accident investigation and modification of the De Havilland Comet. The Hermes was then sold to independent charter airlines for package tour flights from the United Kingdom. It ceased all operations in December 1964.

Variants

HP68 Hermes I	Four Bristol Hercules 101 1,230 kw (1,650 hp). Tailwheel. One built.
HP74 Hermes II	Four Bristol Hercules 121 1,249 kw (1,675 hp). Tailwheel. One built.
HP81 Hermes IV	Four Bristol Hercules 763 1,570 kw (2,100 hp). Tricycle landing gear. Twenty-five built.

HP81 Hermes IVa	Engine redesignated Hercules 773. 100 octane fuel.
HP82 Hermes V	Four Bristol Theseus turboprops 1,860 kw (2,490 hp). Two built.

Hp81 Hermes IV
Crew: (Five)
Max takeoff weight: 37,195 kg (82,000 lb)
Fuel capacity: 14,657 L (3,224 imp gal)
Powerplant: Four Bristol Hercules 763 1,570 kw (2,100 hp)
Propellers: 4-bladed de Havilland Hydromatic, constant-speed

Performance
- Maximum speed: 575 kmh (357mph) at 6,096 m (20,000 ft)
- Cruise speed: 428 kmh (266 mph) max, weak mixture 6,096 m (20,000 ft)
- Range: 5,065 km (3,147 mi) to 5,720 km (3,554 mi), speed dependent
- Service ceiling: 7,500 m (24,500 ft)

Operators: Bahamas (number of airlines 1), Kuwait (1), Lebanon (1), United Kingdom (10)

SARO A36 Lerwick

First Flight: November 1938
Built by: Saunders-Roe Limited, Production: 21
Operational: 1939–1942

The Saunders-Roe A36 Lerwick was a British flying boat designed to work alongside the Short Sunderland in the RAF Coastal Command. Unfortunately, it was quickly discovered that it was a flawed design, and this led to a poor service record and high accident rate.

Development: The Air Ministry issued a specification in 1936 to companies that had experience building flying boats. It was for a medium-range flying boat for reconnaissance, convoy escort,

The Hercules: The Other Engine That Helped Win the War

and anti-submarine duties. The flying boat was to replace the obsolete Saro London and Supermarine Stranraer biplane flying boats. It was to be less than 11,000 kg (25,000 lb) and cruise at 370 kmh (230 mph). Supermarine, Blackburn Aircraft, and Shorts tendered designs. Supermarine won the competition and had an order, but due to Spitfire commitments was unable to fill it. The order was passed to Saunders-Roe. The Lerwick, named after the town, was found to be unstable in the air or on the water. Not only that, but it was also not suitable for 'hands off', stable cruising, which was very important for long range patrols. Modifications were tried including enlarged fin and an increase in the wing angle of incidence. To no avail, the aircraft had a vicious stall and poor rates of roll. Added to this, there were problems with the hydraulics and the structural integrity of the wing floats. The Lerwick was unable to maintain altitude or heading with an engine failure.

Design: The aircraft was a small twin-engined, high wing, all metal cantilever monoplane with a conventional single-stepped, deep, flying boat hull. The hull sides were vertical with up to ten portholes. Entry was by a door on the port side towards the tail. Fuel was in the wings with an overload capability in the hull. The tandem dual controlled cockpit was sufficiently glazed with the engineer sitting aft and starboard of the pilots. Behind the engineer was the sleeping area, kitchen, toilet, and marine storage area. The control surfaces were fabric covered. Tail surfaces were conventional with a cantilevered tailplane and single fin and rudder, all with trim tabs. It had a planning bottom ending in a knife edge beneath the fin. The stabilising floats were attached to the wing by long shock-absorbing V-struts. The bombs or depth charges were carried in two streamlined nacelles behind the engines.

Operational history: A chequered history with No. 209 RAF squadrons due to the aircraft at times being technically grounded due to the high number of accidents, ten, which was half the fleet. The last Lerwicks were delivered in time to see the aircraft withdrawn from front-line service. By 1942 they were all declared obsolete.

Operators Military (Squadrons): RAF (2), one (Coastal) Operational Training Unit, and the Marine Aircraft Experimental Establishment (MAEE). RCAF (2).

Saro Lerwick
Crew: (Six to Nine)
Max takeoff weight: 15,059 kg (33,200 lb)
Fuel capacity: 3,546 L (780 imp gal) normal; 6,546 L (1,440 imp gal) overload
Powerplant: Two Bristol Hercules II 1,011 kw (1,356 hp) at 2,750 rpm at 1,219 m (4,000 ft)
Propellers: 3-bladed De Havilland Hydromatic

Performance
- Maximum speed: 344 kmh (214 mph)
- Cruise speed: 267 kmh (166 mph)
- Range: 2,480 km (1,540 miles) at 322 kmh (200 mph)
- Service ceiling: 4,300 m (14,000 ft)

Armament
Guns: One .303 in (7.7 mm) Vickers K in bow turret, two .303 in (7.7 mm) machine guns in dorsal turret, and four in tail turret. All turrets were powered
Bombs: 907 kg (2,000 lb) or depth charges

Short S26 'G' Class Flying Boat

First Flight: 21 July 1939
Built by: Short Brothers, Production: 3
Operational 1939

The Hercules: The Other Engine That Helped Win the War

The Short S26 G-Class was a large flying boat designed and produced to achieve a non-stop transatlantic capability. Short Brothers wanted to demonstrate their ability to have an aircraft that would achieve long distances. Three years previously they had produced the Short Empire medium range flying boat for the commercial market. It had used the Bristol Pegasus engines. The S26 had been developed at the request of Imperial Airways, but it also had the support of the Air Ministry who was looking at it for long range maritime patrol duties. It resembled the Empire but was much bigger. The start of the Second World War was just weeks away. This resulted in Imperial Airways having the three aircraft for a short time before they were pressed into military service in 1940. They returned to civilian service with British Overseas Airways Corporation two years later.

Development: The success of the Short Empire encouraged Imperial Airways to ask Short to look into a longer-range version so the company would remain competitive. Specifically, they were thinking of a transatlantic route from Foynes, Ireland, to Botwood, Newfoundland, and up to a 4,023-km (2,500-mile) range to carry mail and passengers. Simultaneously, Short were researching a high altitude, pressurised, monoplane aircraft, also for long distance flights. Both designs would be powered by four Bristol Hercules engines. The Short designer, Arthur Gouge, was experimenting with replacing the stepped hull of the Short Empire with flush-fitting extendible hydrofoils. Three flying boats, referred as G-class by Imperial Airways, were ordered. On 27 June 1939 the first S26, the *Golden Hind*, was launched into the water at Shorts' No. 3 shop. A first flight a month later on 21 July was followed by delivery to Imperial Airways on 24 September. The aircraft had performed well with no issues.

Design: The S26 was basically an enlarged Short C-class Empire flying boat. Weight reduction was achieved by extrusions in the structure rather than bent sheet sections. The S26 was designed with flexibility to allow the range to vary as per the number of passengers, twelve for long range to 150 for a 'short hop'. The vast hold was divided by water-tight doors, which were closed for takeoff and landing. It was the largest Short flying-boat built to that time.

Airline/Operational history: Days after the first S26 was accepted by Imperial Airways it and the other two aircraft were seconded to the RAF along with their crews. They immediately flew supplies, such as ammunition and Bristol Beaufighter spare parts, to Gibraltar and the Middle East. The three S26s became X8273, X8274, and X8275, with armour plating in the crew area and inner fuel tanks. In 1941, they joined 119 Squadron RAF for reconnaissance duties followed by 10 Squadron RAAF. In 1945, X8275 travelled to the Belfast factory for a complete overhaul and engine change to the Hercules XIVs. Two aircraft were lost and the first aircraft continued to fly with BOAC between Britain and West Africa until 1947.

Operators Military: RAF (1), RAAF (1), BOAC, MAEE.

Short S26
Crew: (Five) two pilots, navigator, radio operator, flight engineer
Max takeoff weight: 33,339 kg (73,500 lb)
Fuel capacity: 16,366 L (3,600 imp gal)
Powerplant: Four Bristol Hercules IV 1,030 kw (1,380 hp)
Propellers: 3-bladed De Havilland constant speed

Performance
- Maximum speed: 336 kmh (209 mph)
- Cruise speed: 290 kmh (180 mph)
- Range: 5,100 km (3,200 miles)
- Service ceiling: unknown

Armament (military conversion)
Guns: Twelve .303 machine guns, four located in each Boulton Paul BPA Mk II turret Two of the turrets were in the dorsal position and one in the tail

Bombs:
Internal 20 reconnaissance flares, twenty-eight flame floats, and eight smoke floats
External eight 227 kg (500 lb) under the wings

Short S45 Seaford (Sunderland Mk IV)

First Flight: 30 August 1944
Built by: Short Brothers, Production: 10 (2 prototypes)
Operational 1946

The Short S45 Seaford was a long-range lying boat designed as a maritime patrol bomber for the RAF Coastal Command. The Seaford's origins were in the finest flying-boat ever built, the Short Sunderland. The Sunderland would be in production throughout the Second World War and would be in service with the RAF for over twenty years. It would then continue with overseas forces until the 1960s. A remarkable record of reliability and durability not to mention its operational offensive and defensive capability. The Seaford was basically a re-engined Sunderland III with Bristol Hercules engines, in other words a Sunderland IV. It was to be the Sunderland IV but there were too many changes and it was renamed the Seaford.

The Bristol Pegasus-engined flying-boat was an excellent performer with no serious flight problems. Possible spray damage to the inner engine propellers at near maximum takeoff weights was mitigated by advancing the outer engines first until the bow wave was established and then advancing the inner engines to full power.

The Sunderland itself was a development of the Short S23 Empire with its rear step tapered to a knife edge. The two-deck construction featured vertical box frames with watertight bulkheads. The Sunderland was the first British flying-boat to incorporate a manual retractable bow gun turret. Aft of this turret was a marine mooring store, stairs to the upper deck, a toilet, the officers' wardroom – yes, just like on a ship – galley, bomb compartment, crews' quarters, and two beam defensive positions. Behind these were a workshop and the tail turret. The shouldered-mounted wings had Gouge flaps and internal fuel tanks. The flying-boat was an all-metal construction except for the fin and rudder, which were fabric covered aft of their leading edge.

Development: The Air Ministry wanted to improve the performance of the Sunderland patrol aircraft and make it suitable for the Pacific Ocean. This required more powerful engines and enhanced defensive armament. The prototypes had Bristol Hercules XVII engines with the production versions using the Hercules XIX.

Design: It was a high-wing, four-engined, flying boat that stretched the Sunderland III, increased its power, extended and redesigned its planning hull, and had thicker Duralumin skinning on the wing surfaces.

Operational history: The Seaford flew off the River Medway at Rochester for its first flight. It was found to have aerodynamic instability problems which caused the redesign of the fin and tailplane. It was too late to see service in the Second World War. After a two-year operational trial with 201 Squadron RAF six production Seafords were modified at Belfast, Northern Ireland, leased to BOAC with the designation Solent 3.

Variants: Military converted to civilian use.
Operator Military: RAF
Operator Civil: BOAC
Short S45 Seaford
Crew: (Eight to eleven) two pilots, radio operator, navigator, engineer, bomb-aimer, three to five gunners.
Max takeoff weight: 34,019 kg (75,000 lb)
Fuel capacity: unconfirmed, possibly similar to Sunderland 11,602 L (2,550 imp gal)

Powerplant: Four Bristol Hercules XIX 1,280 kw (1,720 hp)
Propellers: 4-bladed, constant-speed

Performance
- Maximum speed: 389 kmh (242 mph) at 150 m (500 ft)
- Cruise speed: 249 kmh (155 mph)
- Fuel: 9,206 L (2,025 imp gal)
- Range: 5,000 km (3,100 mi)
- Service ceiling: 4,300 m (14,000 ft)

Armament
Guns: Six .50 Browning machine guns. Two each in nose, tail, and beam turrets. Two 20 mm Hispano cannons and two fixed .303 Browning machine guns in dorsal turret
Bombs: 2,250 kg (4,960 lb of bombs and depth charges

Short S45A Solent

First Flight: 11 November 1946
Built by: Short Brothers, Production: 16 + 7 converted S45 Seafords
Introduction: 1948

In 1946, BOAC had evaluated the second Short S45 Seaford off the production line and found it wanting. Shorts immediately proposed a version called the S45 Solent 1 to be competitive in the civilian market. It would have three cabins on the lower deck, a promenade, and two cabins on the upper deck. It would provide accommodation for either thirty day or twenty night passengers. The Ministry of Civil Aviation ordered twelve examples for day passengers only, called the Solent 2. They were built at Rochester, Kent, using the jigs of the cancelled Seaford.

The Short S45A Solent was a civilian development of the Short S45 Seaford, which itself was a derivative of the Short Sunderland III. It was the last production model of the large post-war long-distance flying boats. The production line aircraft flew with BOAC and Tasman Empire Airways Limited (TEAL) and later with small airlines, such as Aquila Airways, until 1958.

Design: The Solent was a high-wing monocoque flying boat made of aluminium. Two configurations were available, thirty-six day passengers, or twenty-four day and night passengers. The cabins, yes, the cabins, could be used to seat six or sleep four. Four cabins were on the upper deck and four on the lower. The upper deck included a dining/lounge area and a kitchen. The lower deck had two dressing rooms, toilets, and three freight areas. The Solent 2 had revised accommodation for thirty-four, a dining room, cocktail bar, and a crew of seven. The first of the order flew in November 1946 and the last less than two years later, in April 1948. The type had modifications to its wing floats and struts to increase clearance when taxiing at heavy weights.

Airline history: BOAC operated their Solents on their three times a week Southampton to Johannesburg route down the Nile and East Africa. The journey took four days. The author on a vacation in Kenya visited the Imperial Airways stop on Lake Naivasha, the country's first international airport. This flight lasted until 1950. Meanwhile, BOAC was assuming the Seafords had been declared redundant by the RAF and these became the Solent 3. They were modified for thirty-nine passengers on two decks. By 1949, the Solents were used on the Karachi, Pakistan schedule. Shortly after, in November 1950, BOAC ceased flying-boat operations. Aquila Airways, formed in May 1948, were expanding their flying boat operation with Short Hythe and Sunderland aircraft. TEAL operated five Solents on their Sydney, Fiji, Auckland, and Wellington schedule until 1960. It acquired a Solent 3 and in November 1951 ordered four Solent 4s, a heavier version, with the first flying in May 1949. The Solent 4 carried forty-four passengers with a maximum weight of 36,741 kg (81,000 lb). It used the more powerful 1,521 kw (2,040 hp) Hercules 733 engines which were mounted parallel to the airframe centreline rather than previously toed-out. The range was now 4,828 km (3,000 mi), nearly one and half times the Solent 3. These flying-boats ceased operation on all routes in 1954 except for the Fiji to Tonga service. Trans-Oceanic Airways of Australia bought four ex BOAC Solents in 1951. They were intended for the Australian Sydney, New South Wales to Hobart, Tasmania route. It never transpired and the aircraft passed on to South Pacific Airlines of California, USA. Later Aquila Airways provided service from Southampton to Madeira and the Canary Islands. British flying boat operations ended in 1958.

Variants: Solent II Civilian version of Short Seaford for BOAC, twelve aircraft. Solent III Converted S45 Seaford, seven aircraft. Solent IV Powered by Bristol Hercules 733 engines, four aircraft built.

Operators Civil: BOAC, TEAL (Tasman Empire Airways, Ltd.), Trans Oceanic Airways, Aquila Airways, South Pacific Airlines

Solent II
Crew: (Seven) Two pilots, navigator, radio operator, flight engineer, and two stewards to look after the passengers
Max takeoff weight: 35,380 kg (78,000 lb)
Fuel capacity: unknown
Powerplant: Four Bristol Hercules 637 1,260 kw (1,690 hp)
Propellers: unknown

Performance
- Maximum speed: 439 kmh (273 mph)
- Cruise speed: 393 kmh (244 mph)
- Range: 2,900 km (1,800 miles)
- Service ceiling: 5,200 m (17,000 ft)

Short S29 Stirling

First Flight: 14 May 1939
Built by: Short Brothers, Rochester; Short Bros. and Harland, Belfast, Austin Motor Company, Birmingham, Production: 2,371
Operational 1940

The Short Stirling was the first British four-engine heavy bomber to be introduced into service with the RAF. In the early 1930s, the RAF was interested in developing twin-engine aircraft as it thought that they would be adequate to meet operational demands. However, looking at overseas development, it soon realised that it would need a larger aircraft. The Stirling was designed in response to the Air Ministry Specification 12/36. The Supermarine submission, Type 317, was initially viewed as the primary design with the Shorts S.29 as an alternative. The Type 317 was abandoned and by default the S.29, now known as the Stirling, was ordered into production. Its performance envelope covered two ends of the spectrum. The good was the ability to out-turn the fighters and the bad was the lack of altitude capability. The short-sighted Air Ministry requirement that the aircraft should fit in a 30 m (100 ft) wide RAF 'standard' hangar certainly limited the performance potential of aircraft designed during that time. So much so that the Stirling had a brief, but initially important operational career until 1943, before being relegated to secondary duties. These duties included mining German ports and as invasion glider tugs.

Design: The Stirling was a four-engine heavy bomber specifically designed for strategic bombing. The engines of choice were the Bristol Hercules hung on a mid-mounted wing. It was smaller than pre-war American bombers, but at the time, it was the largest of any British manufacturer. It was bigger than the Handley Page Halifax and Avro Lancaster which would follow. This was because these aircraft were originally designed as twin-engine aircraft.

The Hercules: The Other Engine That Helped Win the War

The Stirling was also different in that it was designed for two pilots housed in a glazed cockpit abeam the forward end of the bomb cells; the flight engineer had his own separate station. The bomb aimer, with drift sight, camera, and control over the aircraft's autopilot, would conduct his duties in a prone position underneath the front gun turret. Just aft of the wireless operator's position was a rest and oxygen tank storage area. Further aft was a large open storage area for flame floats and reconnaissance flares, with a ventral turret in early models, and which included a toilet, port side entry door, and rear turret. The mid-mounted wing, similar to earlier Short flying boats, was a cantilevered, two-spar construction clad in flush-rivetted aluminium alloy sheeting. It housed three large self-sealing fuel tanks contained within the two main spars. A fourth fuel tank was contained within the leading edge of the wing root. Additional fuel could be carried as necessary in six ferry tanks in the wing bomb cells. Gouge type flaps were incorporated, similar to the flying boat design. The fuselage consisted of four sections bolted together with continuous stringers through each section. The bomb section was comprised of three parallel bomb cells restricted in size to 227 kg (500 lb) and 907 kg (2,000 lb) bombs only. This would limit its operational capability as larger bombs, the 1,800 kg (3,097 lb)), and special bombs were designed and became available. Hydraulic power was used to control the throttles, a source of irritation to the crew with its sometimes slow responsiveness. The Bristol Hercules engines were upgraded from the Mk II, to VI, to XI, and finally the XVI.

Development: The Air Ministry ordered two full-scale S.29 prototypes. However, Shorts constructed the S.31 half-scale model to test the feasibility of the design; they had previously taken the same approach for some flying boats. Landing gear was lengthened to increase wing incidence and takeoff performance. First flight was on 19 September 1938. The Stirling had twin tailwheels. The impending threat of war meant that the Air Ministry order went from 100 to 1,500 seemingly overnight.

The Hercules Powered Aircraft

Up to twenty factories were involved with the production of the Stirling, and by May 1940 the first production Stirling went airborne. The whole manufacturing process was very slow at this time due both to organisational problems and bombing attacks on the main factories. Satellite facilities would be established to counter this offensive and keep production going. A high altitude, Bristol Hercules 17 SM engine version, a Canadian-built aircraft using American Wright engines, a civil version, and a version with Bristol Centaurus engines were all proposed but never built.

Operational history: In August 1940 the first Stirling was delivered to No 7 RAF Squadron RAF at RAF Leeming, North Yorkshire. By January 1941 it had achieved operational status and on 10/11 February took part in its first operational mission. Three Stirlings were assigned a bombing mission to destroy the fuel storage tanks at Vlaardingen, near Rotterdam, Holland. It was deemed a success. During its first year of service three squadrons were equipped with Stirlings and 150 had been produced. They flew in day and night missions and, in cooperation with fighters, formed a tactical fighting group referred to as the 'Circus Offensive'. It also played a pioneering role in the formation of the new Pathfinder Squadrons, developing specialist navigation equipment and procedures, and target-finding squadrons to assist the Main Force.

By May 1943, they were part of one hundred strong Stirling bomber raids and also part of the famous 1,000 bomber raids. The Stirling had demonstrated that by its construction it could take severe damage and still make it home with its crew to fight another day. Despite its lack of ability to climb above 5,000 m (16,500 ft) it could still manoeuvre well against the attacking fighters due its thick wing aerofoil construction. In order to have the range to reach targets deep in Germany or Italy, the bomb load had to be restricted to a quarter of its maximum bomb load; this was the same as a medium bomber such as the Vickers Wellington. The Stirling aircraft handled well compared to some of the other heavy bombers. However, the lower ceiling made it more vulnerable than the other heavy bombers flying above it in mixed type formations. So, in spite of its attributes, the Stirling loss rate was unacceptably high compared to other bombers, eighty per cent lost in the first six months of service.

Flying characteristics: Airborne, the Stirling was very manoeuvrable for an aircraft of that size. However, like other large, multi-engine, tail-wheeled aircraft, it was a challenge during takeoff and landing. This was compounded by the fact that due to the urgent demands of war, pilots were often inexperienced, lacking in total flying time. During takeoff it was easy to lose directional control and enter a ground-loop, possibly causing extensive damage. The RAF instituted a special training and certification course for Stirling pilots. A technique of initially slowly feeding in the power from the right engines until the rudder became effective to maintain directional control was introduced. As mentioned previously, the hydraulic throttles compounded the problem. Landings presented a problem with the tendency of the aircraft to suddenly lose lift at slow speeds and stall, causing extensive damage.

Variants

Model	Description
S.31 (M4)	Half scale flying test bed powered by four Pobjoy Niagara 7 engines
Stirling I	Bristol Hercules II, III, XI engines
Stirling II	Wright R-2600 Twin Cyclone engines, Canada prototypes, no production
Stirling III	Bristol Hercules VI, XVI engines, electronic counter-measures
Stirling IV	Bristol Hercules VI, XVI engines, para-dropping, glider towing, assault transport. (Battle of Normandy, Operation *Market Garden*)
Stirling V	Bristol Hercules VI, XVI engines, cargo aircraft

Operators Military (Squadrons): Egyptian Air Force; Germany, Luftwaffe; United Kingdom, Royal Air Force (28, including No. 75 New Zealand Squadron)

Operations Civil: Belgium, post-war civilian transport

Short Stirling I SRS II, III
Crew: (Seven) two pilots, flight engineer, navigator/bomb aimer, wireless operator/front gunner, two air gunners
Max takeoff weight: 31,751 kg (70,000 lb)
Fuel capacity: 10,247 L (2,254 imp gal)
Powerplant: Four Bristol Hercules XI 1,100 kw (1,500 hp)
Propellers: 3-bladed metal, fully feathering, constant-speed propeller

Performance
- Maximum speed: 454 kmh (282 mph) at 3,800 m (12,500ft)
- Cruise speed: 320 kmh (200 mph)
- Range: 3,750 km (2,330 miles)
- Service ceiling: 5,000 m (16,500 ft)

Armament
Guns: Eight .303 in (7.7 mm) Browning machine guns, two in powered nose turret, four in tail turret, two in dorsal turret.
Bombs: 6,350 kg (14,000 lb) maximum

Vickers Valetta

First Flight 30 June 1947
Built by: Vickers-Armstrongs Ltd, Production: 262
Operational 1948

The Valetta was a twin-engine military transport aircraft developed by Vickers from the Vickers VC 1 Viking civil transport. Post-war, the RAF had a requirement for a transport to fill multi-roles including troop transport, air ambulance, glider tug, paratroop carrier, supply drop, and freighter. It was important to have flexibility to meet these roles with the result that the furnishings were easily removed and could be exchanged for new configurations. It would require the more powerful Bristol Hercules 230 model, a reinforced floor and strengthened landing gear. It formed a major part of the post-war transport component and saw service in the Suez Crisis, Aden, and the Malayan Emergency.

Development: The Valetta was developed from the Viking which itself was developed from the Vickers Wellington bomber. The Viking had gone through extensive testing with the RAF before the Valetta was accepted into the RAF Transport Command. The aircraft would replace many transports left over from the Second World War. The aircraft would also be developed into a VIP transport and training aircraft and ultimately into the Vickers Varsity, a nosewheel tricycle landing gear version.

Design: The Valetta was an all-metal mid-wing monoplane with a tailwheel landing gear configuration. It had two large, reinforced cargo doors in the fuselage to allow carriage of larger pieces of cargo. The floor could sustain 680 kg (1,500 lb) point contact weight, had floor rollers, and had anchoring attachments to secure the cargo. Interior furnishings included different styles and arrangements of seating, a loading winch, vehicle ramps, soundproofing insulation, and additional fuel tanks. It had a variable capacity of thirty-four fully equipped soldiers, or twenty stretchers with two medical attendants, or twenty paratroopers with nine 159 kg (350 lb) air-droppable supply containers. The external fuselage was modified to carry loads on twin racks and the rear fuselage had a pyramid structure as an anchoring point and release mechanism for towing gliders.

Operational history: The Valetta replaced the Douglas Dakota and was used by RAF Transport Command over active conflict zones, such as parachute drops during the 1956 Suez Crisis.

Variants: C1 transport, C2 VIP transport, T3 navigation trainer, T4 airborne intercept role.

Operators Military (Squadrons): RAF (18), Navigation School (2), RAF College.

Vickers Valetta C1
Crew: (Four)
Max takeoff weight: 16,556 kg (36,500 lb)
Fuel capacity: 3,290 L (724 imp gal)
Powerplant: Two Bristol Hercules 230 1,500 kw (2,000 hp)
Propellers: 4-bladed de Havilland/Rotol four-bladed, constant-speed

Performance
- Maximum speed: 415 kmh (258 mph) at 3,000 m (10,000 ft)
- Cruise speed: 277 kmh (172 mph)
- Range: 2,350 km (1,460 miles)
- Service ceiling: 6,600 m (21,500 ft)

Vickers Varsity

First Flight: 17 July 1949
Built by: Vickers-Armstrongs, Production: 163
Operational 1951–1976

The Vickers Varsity was a British twin-engine trainer in service with the RAF. It was designed to replace the Wellington T10, Valetta T3, and Valetta T4.

Development: The Varsity, derived from the Viking and although similar to the Valetta, had wider-span wings, a longer fuselage, and a tricycle landing gear. It was equipped with a distinctive under-fuselage pannier for accommodating a bomb-aimer and up to 24 practice bombs. The trainee would lie in a prone position to release bombs over the target.

The Hercules: The Other Engine That Helped Win the War

Operational history: The first production aircraft were used to train pilots on multi-engine aircraft at RAF Swinderby

Variants: T1, Tp 82 (Sweden)

Operators Military (Squadrons): Royal Jordanian Air Force, Swedish Air Force, United Kingdom: RAF (10), Bomber Command Bombing School, Central Navigation and Control School, Royal Air Force Cranwell, Central Flying School, Flying Training School (3), No. 201 Advanced Training School, Air Navigation School (4), Air Electronics School, Radio School, Royal Aircraft Establishment, Empire Test Pilot's School, Aeroplane & Armament Experimental Establishment.

Vickers Varsity T Mk 1
Crew: (Four)
Max takeoff weight: 17,010 kg (37,500 lb)
Fuel capacity: 3,369 L (741 imp gal)
Powerplant: Two Bristol Hercules 264 1,450 kw (1,950 hp)
Propellers: 3-bladed de Havilland Hydromatic

Performance
- Maximum speed: 463 kmh (288 mph) at 3,000 m (1,450ft)
- Cruise speed: 385 kmh (239 mph)
- Range: 4,262 km (2,648 miles)
- Service ceiling: 8,700 m (28,700 ft)

Armament
Bombs: 270 kg (600 lb) total practice bombs, 11.3 kg (25 lb) each in external pannier

Vickers C1 Viking

First Flight: 22 June 1945
Built by: Vickers Armstrongs Production: 163

The Hercules Powered Aircraft

Introduction: 1946

The Vickers Viking is a British twin-engine short-range airliner designed in the latter stages of the Second World War and derived from the famous Vickers Wellington bomber. The Viking would fill the niche of the last piston-era transport aircraft before the advent of the Vickers Viscount turboprop's first flight in 1948. Its airframe was also used for the first flight of a pure jet transport, an experimental version fitted with Rolls-Royce Nene engines.

Design/Development: In 1944, the Ministry of Aircraft Production ordered three prototypes from Vickers for a 'Wellington Transport Aircraft' as an interim design until the more advanced designs specified by the Brabazon Committee could be developed. The wing and landing gear were pure Wellington, but the fuselage was the newly designed component. Following successful trials of the prototypes, fifty production aircraft were ordered. British Overseas Airways Corporation (BOAC) flew the Viking in spring 1946. The prototypes had further trials with the RAF, resulting in a military version, Viking C2, and the modified Valetta C1. The initial production batch, later designated Viking 1A, carried twenty-one passengers. It had a metal fuselage but the wings, outboard of the engines, and tailplane were fabric-covered geodetic structures. The Viking 1 was of normal stressed metal construction. The Viking 1B was 71 cm (28 in) longer with uprated Bristol Hercules engines. On 6 April 1948, it was this version that had the Nene engines installed. This aircraft flew to Paris on 25 July 1948, the 39th anniversary of Bleriot's crossing the English Channel by powered aircraft for the first time. The flight took only thirty-four minutes.

Airline history: Initially transferred to BOAC, nine Vikings were then transferred to British European Airways (BEA), a new airline for European routes. The VC, Vickers Commercial, designation was soon replaced by company designations, an example being Type 615 for the Argentine government. BEA had the seating increased to thirty-six and operated them until 1954 when they were replaced by another Vickers product, the turboprop Viscount. The Vikings then were operated by independent airlines and charter airlines after the mid 1950s.

Variants

Model	Description
Viking	Prototypes, two Bristol Hercules 130 engines 1,250 kw (1,675 hp)
1A	Initial production, Bristol Hercules 630 engines 1,261 kw (1,690 hp)
1	Main production, Bristol Hercules 630 engines 1,261 kw (1,690 hp)
1B	Long nose version
Nene Viking	Trials with two Rolls-Royce Nene 22.3 kN (5,000 lbf)
C2	Military version, VIP, RAF King's Flight

Operators Civil: Argentine (3), Austria (1), Belgium (1), Denmark (1), Egypt (1), France (6), Germany (7), India (3), Iraq (2), Ireland (1), Kuwait (1), Mexico (2), Portugal (1), Pakistan (1), South Africa (5), Southern Rhodesia (1), Switzerland (1), Trinidad & Tobago (1), United Kingdom (35)

Operators Military: Argentine, Australia, Jordan, Pakistan, United Kingdom

Viking 1B
Crew: (Four)
Max takeoff weight: 15,422 kg (34,000 lb)
Fuel capacity: 3,400 L (740 imp gal)
Powerplant: Two Bristol Hercules 634 1,260 kw (1,690 hp)
Propellers: 4-bladed de Havilland/Rotol constant speed

Performance
- Maximum speed: 423 kmh (263 mph) at 300 m (1,000 ft)
- Cruise speed: 340 kmh (210 mph)
- Range: 2,700 km (1,700 miles)
- Service ceiling: 7,600 m (25,000 ft)

Vickers Wellesley

First Flight: 19 June 1935
Built by: Vickers-Armstrongs, Production: 177
Operational: 1937

The Vickers Wellesley, as well as the next aircraft listed, was named after Arthur Wellesley, 1st Duke of Wellington. It was produced by the British aircraft manufacturer Vickers-Armstrongs at Brooklands (Weybridge), Surrey, developed as a monoplane from the Type 253, geodesic airframed biplane. Initially, the biplane was ordered but as the industry quickly changed it was re-ordered as the Type 256 monoplane after it had successfully flight tested. In 1938, a demonstration flight was conducted by three specially prepared Wellesleys that flew non-stop for two days completing an Ismailia, Egypt, to Darwin, Australia, flight. The type was unsuited for the European war but did see service overseas in the desert theatres of East Africa, Egypt, and the Middle East.

Design: It was a single engine monoplane with a very high aspect ratio wing. It had manually operated retractable landing gear; it was, after all, 1935. The wing was redesigned a few times to add strength. The bomb container doors were eliminated as they were causing vibration and the bombs were carried in open geodesic panniers underneath the wings.

Development: Vickers were determined to make the transition from airships to aircraft and secure orders for fixed wing aircraft. These designs were concentrated around large-engined biplanes. Barnes Wallis was involved with some of the proposals. The Type 253 was in response to a requirement for a general purpose aircraft to carry out level bombing, army co-operation, reconnaissance, dive and torpedo bombing, and evacuation of casualties. It was awarded a development contract as the biplane, but Vickers were working on a monoplane design that would fulfil the same requirements. The Type 253 biplane, after testing against other manufacturer's submissions, was selected and awarded a contract for 150 aircraft. Before starting production, it was changed to the successful design of the Type 246 monoplane after extensive metallurgical and structural testing. In January 1937, the Wellesley performed its maiden flight.

Operational history: The Wellesley was in service as a bomber by April 1937 at Finningley, South Yorkshire, with No.76 RAF Squadron. It was quickly replaced by Handley Page Hampdens, Armstrong Whitworth Whitleys, and Vickers Wellingtons by the start of the war. The overseas service included operations against Italian Forces in Eritrea, Ethiopia, and Somaliland, and bombing raids flown from Aden to Addis Ababa. After maritime reconnaissance over the Red Sea, the Wellesley's active service came to an end in 1942. The design principles of the Wellesley would serve Vickers well as it designed its famous Vickers Wellington.

Variants

Type	Description
281	Company designation
287 Mk I	Two, later three seat, aircraft. Separate canopies.
287 Mk II	Extended canopy, pilot and bomb-aimer.
289	Engine test bed, Bristol Hercules HE15.
291	Blind flying model.
292	Five long-distance, record-breaking aircraft.
294	Strengthened wing for cutting barrage balloon cables.
402	Three-seat experimental aircraft

Operators Military (Squadrons): Royal Air Force (12), South African Air Force

Vickers 689 Wellesley Bristol Taurus version
Crew: (Three) Pilot, bomb-aimer, rear gunner.
Max takeoff weight: 5,670 kg (12,500 lb)
Fuel capacity: unknown
Powerplant: One Bristol Hercules I 1,007 kw (1,350 hp), HE15 engine test only
Propellers: unknown

Performance
- Maximum speed: 367 kmh (228 mph) at 6,000 m (19,700 ft)
- Cruise speed: 290 kmh (180 mph)
- Range: 1,960 km (1,220 miles)
- Service ceiling: 7,800 m (25,500 ft)

Armament
Guns: one 7.7 mm (.303 in) Vickers machine gun right wing, one 7.7 mm (.303 in) Vickers K machine gun in rear cockpit.
Bombs: 910 kg (2,000 lb)

Vickers Wellington

First Flight: 15 June 1936
Built by: Vickers-Armstrongs, Total Production: 11,461 (11,462?)
Operational: October 1938

The Wellington was designed by Vickers chief designer Rex Pierson during the mid-1930s. It had a unique geodetic airframe fuselage structure. Vickers designed the Wellington in response to an Air Ministry specification for a twin-engine day bomber delivering new higher performance criteria. Its competition was the Armstrong Whitworth Whitley and Handley Page Hampden. In the early years of the war, it was also used as a

The Hercules: The Other Engine That Helped Win the War

night bomber, becoming one of the principal aircraft of Bomber Command. Although superseded by the four-engine 'heavies' in 1943, it continued service in other capacities, such as an anti-submarine aircraft. The Wellington holds the distinction of being the only bomber to have been produced for the whole war and having been produced in greater numbers, over 11,000, than any other aircraft. It went on to be developed into the maritime reconnaissance Vickers Warwick and many parts were used in the civilian transport VC.1 Viking.

Development: In 1932, Vickers chief structures designer Barnes Wallis proposed to continue with the geodetic airframe of the Wellesley for the new bomber. The structure demonstrated superior strength during testing at Royal Aircraft Establishment, Farnborough. Vickers studied the application of air and liquid-cooled engines to the airframe. Initially, the engines chosen were the air-cooled Bristol Pegasus and liquid-cooled Rolls-Royce Goshawk. The air-cooled engine was the final selection, Bristol Pegasus or the sleeve-valve Bristol Perseus. Other production features were the variable-pitch propellers, Vickers nose and tail guns, spring-loaded bomb doors, and a major change to a mid-wing mounting. Weights and bomb loads were continually revised upwards as the looming war hastened development of a bigger, faster bomber. In 1936, RAF Bomber Command was formed.

In June 1936, the Wellington made its first flight. Major redesign happened rapidly, with a deepened fuselage, extended nose, and increased crew complement. Other changes were constant-speed propellers, retractable tailwheel, and Nash & Thompson ventral gun turret. By December 1937, the first of many marks of production aircraft made its first flight. The orders rapidly increased, including the Mk II with Rolls-Royce Merlin engines. So did the production rates; by 1942, Broughton, North Wales, produced 130 aircraft per month and in 1943, for propaganda purposes, a Wellington was built in twenty-three hours and fifty minutes.

The Hercules Powered Aircraft

Design: the Wellington was designed as a twin-engine, long-range bomber, initially with Bristol Pegasus engines driving De Havilland two-pitch propellers. Various combinations of engines and propellers were tried and used. This included the Bristol Hercules and the iconic Rolls-Royce engine. The Wellington had some easily recognisable features, namely the high aspect ratio of its tapered wing, the tall, single, vertical stabiliser, and the depth of its fuselage.

It typically had a crew of five, with the bomb aimer located in the nose. It was continually developed during the war to allow different sizes and loads of bombs to be carried. It could be fitted with dual controls to allow training on type during conversion training. There was a nose turret, a tail turret, and a retractable ventral turret of .303 in (7mm) Browning machine guns. These turrets were the enclosed type manufactured by Nash & Thompson, which replaced the original Vickers designed units.

Barnes Wallis, the inventor of the 'bouncing bomb' Tallboy and Grand Slam bombs, among other things, designed and invented a unique geodesic fuselage structure. The Wellington had one of the most robust airframes ever developed, and pictures of its skeleton largely shot away, but still sound enough to bring its crew home safely, are still impressive. The fuselage consisted of 1,650 elements of duralumin W-beams formed into a metal framework with wooden battens attached to the beams. Irish linen was then attached and treated with layers of dope to form the skin of the aircraft. This construction allowed the flexibility to adapt to increased weight, carrying ever-larger bombs, and the addition of long-range fuel tanks. This flexibility would continue for the production life of the aircraft. The specialised geodetic construction did not lend itself to other construction methods and hence was not easily transferred to other manufacturing companies. The reverse was also true, as Vickers would find it very hard to adapt to other construction methods.

The Hercules: The Other Engine That Helped Win the War

Hercules	Mark
III	Vickers 299, 417 Wellington III
VI	Vickers 454, 458 Wellington XI
VI	Vickers 455 Wellington XII
VIII	Vickers 407, 426 Wellington V
X	Vickers 417 Wellington III
XI	Vickers 417 Wellington III
XI	Vickers 428 Wellington DWI Mk III
XI	Vickers 440 Wellington X
XVMT	Vickers Wellington V
XVI	Vickers 437 Wellington X
XVI	Vickers 439, 440 Wellington IX, X
XVI	Vickers 478, 487 etc Wellington X, XI, XII, XIII, XIV, XVII, XVIII
XVII	Vickers 466 etc Wellington X, XI, XII, XIII, XIV, XVII, XVIII
100	Vickers 478 Wellington X

Operational history: On 4 September 1939, Wellingtons performed the first RAF bombing raid of the war, targeting German shipping at Brunsbuttel. Initially, the focus was on daylight attacks on German shipping, but it was too costly in aircraft lost to enemy fighters, so within a year the Wellington was changed to night operations against industrial targets. Coastal Command also adopted the Wellington and used it in anti-submarine duties. Superseded in the European theatre, it went on to serve in the Middle and Far East. In 1944, a specially equipped aircraft was used as an airborne radar early warning and control aircraft for Beaufighters and Mosquitos attacking enemy bombers and V1 rockets.

Variants

Type	Description and Number Built
RAF	
TYPE 271	Prototype
TYPE 285 Mark I	Pre-production prototype, Bristol Pegasus engines
TYPE 290 Mark I	Production version, Bristol Pegasus XVIII, 183

The Hercules Powered Aircraft

Type	Description and Number Built
TYPE 408 Mark IA	Production version, Bristol Pegasus XVIII, 187
TYPE 416 Mark IC	Main Bristol Pegasus version, six crew, 2,685
TYPE 406 Mark II	Identical to IC with Rolls-Royce X engines, 401
TYPE 417 B Mark III	Bristol Hercules III/XI version, 1,517
TYPE 424 B Mark IV	Production version Pratt & Whitney Twin Wasp engines, 220
TYPE 442 B Mark VI	Rolls-Royce Merlin R6SM, high altitude, pressurised, 63
TYPE 440 B Mark X	Main Bristol Hercules version, XVIII, 3,804
COASTAL COMMAND	
TYPE 429 GR Mark VIII	Mark IC conversion for Coastal Command, 307
TYPE 458 GR Mark XI	Maritime version of B Mark X, ASV Mark II radar, 180
TYPE 455 GR Mark XII	Maritime version of B Mark X, torpedoes, radar Mark III, 58
TYPE 466 GR Mark XIII	Maritime version of B Mark X, nose turret, no waist guns, 844
TYPE 467 GR Mark XIV	Maritime version of B Mark X, RP-3 rocket rails, 841
TRANSPORT COMMAND	
TYPE C Mark XV	Mark IA conversion, 18 troops
TYPE C Mark XVI	Mark IC conversion, 18 troops
TRAINING COMMAND	
TYPE 487 T Mark XVII	Bristol Hercules XVII engines, air intercept radar
TYPE 490 T Mark XVIII	Production with Bristol Hercules XVI engines, 80
TYPE T Mark XIX	Conversion of Mark X, in service until 1953
TYPE 619 T Mark X	Navigation training version

The Hercules: The Other Engine That Helped Win the War

Type	Description and Number Built
EXPERIMENTAL/CONVERSIONS	
TYPE 298 Mark II	Prototype Rolls-Royce Merlin X engines
TYPE 299 Mark III	Prototype, two only
TYPE 410 Mark IV	Prototype Pratt & Whitney Twin Wasp engines
TYPE 416 (II)	Original Wellington II converted with Vickers S in dorsal gun
TYPE 418 DWI Mark I	Directional Wireless Installation minesweeper, 4
TYPE 419 DWI Mark II	DWI with increased generator power, DH Gipsy engine, 11
TYPE 407 Mark V	Bristol Hercules VIII, turbocharged for high altitude, 1
TYPE 421 Mark V	Bristol Hercules VIII, turbocharged for high altitude, 1
TYPE 431 Mark VI	Mark V conversion for high altitude, Rolls-Royce Merlin 60 srs
TYPE 449 Mark VIG	Production version of Type 431, 2
TYPE 430 Mark VII	Test bed for 40 mm Vickers S gun turret
TYPE 435 Mark IC	Conversion for turbinlite test
TYPE 437 Mark IX	Mark IC conversion for troops
TYPE 439 Mark II	Conversion for 40 mm Vickers S gun in nose turret, 1
TYPE 443 Mark V	Test aircraft for Bristol Hercules VIII engines
TYPE 445 (I)	Test aircraft for tail mounted Whittle W2B/23 turbojet, 1
TYPE 454 Mark IX	Bristol Hercules VI/XVI with ASV Mk II/III radars
TYPE 459 Mark IX	Bristol Hercules VI/XVI with ASV Mk II/III radars
TYPE 470	Whittle W2B turbojet
TYPE 486	Whittle W2/700

The Hercules Powered Aircraft

Type	Description and Number Built
TYPE 478 Mark X	Test aircraft for Bristol Hercules 100 engine, 1
TYPE 602 Mark X	Test aircraft for two Rolls-Royce Dart turboprop engines
TYPE 417 Mark III	Glider testing, Hadrian, Hotspur, and Horsa

Operators Military (Squadrons): Royal Australian Air Force (3), Royal Canadian Air Force (13), Czechoslovakian Air Force (exile Great Britain) (1), Free French Air Force (2), France (Aeronavale) (2), Nazi Germany (1 captured aircraft), Hellenic Air Force (2), Royal New Zealand Air Force (1), Polish Air Force (exile Great Britain) (4), Portuguese Air Force (1 aircraft interned), South African Air Force (3), United Kingdom, Royal Air Force (42 units/55 squadrons), United Kingdom, Fleet Air Arm (6)

Mark X
Crew: (Five) pilot, navigator, bomb aimer/gunner, two air gunners
Max takeoff weight: 16,556 kg (36,500 lb)
Fuel capacity: unknown
Powerplant: Two Bristol Hercules VI/XVI 1,249 kw (1,675 hp)
Propellers: 3-bladed de Havilland Hydromatic

Performance
- Maximum speed: 410 kmh (255 mph)
- Cruise speed: unknown
- Range: 3,034 km (1,885 miles)
- Service ceiling: 6,706 m (22,000 ft)

Armament
Guns: 6-8 .303 in (7.7 mm) Browning machine guns, 2 in tail, 2 in nose, 2 in waist. 4 in tail and waist in some models
Bombs: 1,814 kg (4,000 lb)

FOREIGN AIRCRAFT

Breguet 890H Mercure (Mars)

First Flight: 1 March 1941
Built by: Breguet Aviation, Production: 1 prototype, Total Models: 3 prototypes

The Mercure was a Breguet proposal for a medium-capacity military and civilian transport to replace the Douglas C-47. The civilian version was called the 890H Mercure and the military version the 891R Mars.

Development: The Mars had room for twenty paratroopers and had parachute doors on each side, a floor chute for dropping containers, and a towing hook for a glider.

Design: It was an all-metal high-wing cantilever monoplane with tricycle landing gear and an innovative swing tail and loading ramp for easy access to the fuselage for easy cargo loading.

Variants: The Mercure 890H had Bristol Hercules 738 engines. The Mercure 892S had four Renault 12S engines, cargo door on the starboard fuselage, and two passenger doors on the port side. It was designed to seat forty passengers. The 891R Mars had two Gnome-Rhone 14R-2000 engines. One variant of each type was built and did not go into production.

890H Mercure
Crew: One plus
Max takeoff weight: 16,000 kg (35,274 lb)
Fuel capacity: unknown
Powerplant: Two Bristol Hercules 738 1,521 kw (2,040 hp)
Propellers: unknown

Performance
- Maximum speed: 350 kmh 220 (mph) at 2,500 m (8,200 ft)
- Cruise speed: 285 kmh (177 mph)
- Range: 1,000 km (620 miles)
- Service ceiling: unknown

CASA C-207 Azor

First Flight: 28 September 1955
Built by: CASA, Production: 22, including two prototypes
Operational 1960–1980s

The CASA C-207 Azor was a scaled-up version of the CASA 202 Halcon and was initially designed for the domestic civil market. It was purpose-built to replace the existing transports in service at the time. These were Spanish-built German designed aircraft, the CASA 2-111 (Heinkel HE 111) and CASA 352 (Junkers Ju 52).

Development: It did not receive any orders; it was unpressurised, lacking in performance and capacity compared to its European and American competition, so the government supported the aircraft by purchasing ten for the Spanish Air Force. It was designated the T7A. This was followed by ten T7B aircraft.

Design: A low-wing, tricycle-gear monoplane with two engines.

Operational history: 1960–1980s

Variants (Number built): A: (2) military, B: (10), two fitted with Pratt & Whitney Double Wasp engines, C: large cargo door at rear of fuselage and space for thirty-seven paratroopers.

Operators Military: Spanish Air Force

CASA 207B
Crew: Four
Capacity: 40 passengers/400 kg cargo
Max takeoff weight: 16,500 kg (35,640 lb)
Fuel capacity: unknown
Powerplant: Two Bristol Hercules 730 1,522 kw (2,040 hp)
Propellers: unknown

Performance
- Maximum speed: 420 kmh (261 mph)
- Cruise speed: 400 kmh (249 mph)
- Range: 2,350 km (1,460 miles)
- Service ceiling: 8,000 m (26,250 ft)

Fokker T.IX

First Flight: 11 September 1939
Built by: Fokker, Production: 1

The T.IX was a Dutch bomber designed and built for the Royal Netherlands East Indies Army Air Force as a replacement for their obsolete bombers.

Development: It was started in 1938 as the company's first all-metal bomber project. It first flew in 1939 but was damaged in April 1940 when it collided with a hangar door. The subsequent invasion by Germany halted all further development.

Design: The T.IX was a mid-wing monoplane with retractable gear and twin fins and rudders

T.IX
Crew: One plus
Max takeoff weight: 11,200 kg (24,692 lb)
Fuel capacity: unknown

Powerplant: Two Bristol Hercules II 1.025 kw (1,375 hp)
Propellers: unknown

Performance
- Maximum speed: 440 kmh 270 (mph)
- Cruise speed: unknown
- Range: 2,720 km (1,690 miles)
- Service ceiling: 8,000 m (26,000 ft)

Armament
Guns: intended one 20 mm cannon, twin 12.7 mm dorsal and ventral machine guns
Bombs: Internal up to 2,000 kg (4,409 lb)

Nord Noratlas

First Flight: 10 September 1949
Built by: Nord Aviation, Production: 425 (1949–1961)
Operational: 6 September 1953–1989

The Nord Noratlas was a dedicated military transport aircraft developed and produced by Nord Aviation, a French aircraft manufacturer. The fundamental idea in the late 1940s was to replace the aging Second World War aircraft that had been in service with the Armee de l'Air of France during and after the war. In response to a government Direction Technique Industrielle (DTI), competition, Nord produced a Nord 2500 proposal, which was the design chosen. However, the aircraft, with Gnome-Rhone 14R engines, did not meet requirements and it was revised and released as the Nord 2501. This aircraft used the SNECMA built under licence Bristol Hercules 738 and 739 engines and met all standards. It went into production and was introduced into service with the Armee de l'Air in December 1953. The aircraft then found other air forces as customers and some, the Israeli Air Force, the Hellenic Air Force, and

The Hercules: The Other Engine That Helped Win the War

the Portuguese Air Force operated the Noratlas under combat conditions, a sure testament to the aircraft's and the Bristol Hercules engine's success. The aircraft was produced over ten years and was adapted to a variety of roles as the need arose. There were a few civil operators of the type.

Development: The French Air Force was left with two primary transport aircraft at the end of the Second World War, the Junkers Ju 52 and the Douglas C-47. The Ju 52 was made in France for some time after the end of the conflict. The C-47 came from United States war surplus supplies. Both aircraft had admirable service records but did not have the capabilities of the newer designs of cargo aircraft. For one, they were both tail wheel aircraft, which made them awkward to load through the cargo side doors, especially heavy or unusually shaped cargo. The newer cargo aircraft had the ability to carry larger payloads. The DTI wanted a medium-weight aircraft that would offer operational flexibility. Several companies in addition to Nord submitted proposals. Breguet offered the BR-89R Mars, a swing loading tail, high-wing, two-engine aircraft with tricycle landing gear. It had two side parachute doors, a floor chute for dropping containers, and a glider towing hook. Sud-Ouest offered the SO-30C design.

In 1949, the first Noratlas made its maiden flight with Gnome-Rhone-14R engines of 1,193 kw (1,600 hp) driving three-bladed, variable-pitch propellers. Flight testing revealed that the aircraft was underpowered. To solve this problem a licensing arrangement was made with the Bristol Aeroplane Company to build their Hercules engine. The second prototype installed Hercules 738 and 739 engines capable of producing 1,521 kw (2,040 hp) driving a four-bladed Rotol propeller and was rechristened the Nord 2501. Suitably impressed by its performance, DTI ordered three more aircraft to the same Nord 2501 standard. These three were flown against the Fairchild C-82 Packet for comparison purposes and found to have superior performance. In July 1951, a production order for thirty-four aircraft was issued.

Unfortunately, tragedy struck the Nord 2501 programme as the prototype was destroyed, along with the crew, in a flight test accident. In January 1953 the aircraft was renamed Nord Noratlas by the widow of the pilot killed in the crash.

In spite of the setback, Nord went on to produce the first thirty-four aircraft on schedule and completed a total order of 228 aircraft for the French Air Force. Several different models were considered but never came to production. One unique civilian model that was built was the Nord 2502. It featured wing-tip mounted Turbomeca Marbore turbojets, giving improved takeoff performance. They were not used in flight. Another was a proposed STOL version with airbrakes, larger flaps, and height adjustable landing gear.

Design: The Nord Noratlas was a French twin-boom, twin-engine, shoulder-wing military aircraft. It had a pod-like fuselage slung in between the two tail booms, which were attached to the wings directly behind the engines. They formed part of the engine nacelle rear fairing. This arrangement allowed easy loading of cargo into the fuselage through the tail clamshell doors, as the empennage was mounted higher on the booms. The shoulder-mounted wings were attached to the fuselage. The aircraft was deliberately designed to operate in challenging conditions such as on unprepared surfaces. The widely spaced landing gear, with special low-pressure tires, gave additional stability when operating on rough terrain, such as in desert operations. The propellers had sufficient clearance to operate on any kind of surface.

Operational history: The French Air Force used the Noratlas predominantly in a cargo-carrying role. Ten aircraft had been converted to the dual role of passenger and cargo. In 1962, at the end of the Algerian War of Independence and associated operations, some of the aircraft were converted to perform secondary service. For instance, the most successful secondary use of the aircraft was as a platform for electronic warfare.

The Hercules: The Other Engine That Helped Win the War

It served the longest in this capacity and was the last variant to cease operations in 1989. The Noratlas was famous among the French public as it was remembered for its role in the Suez Crisis of 1956. It air-dropped paratroopers south of Port Said and Port Fouad in Egypt during a quick-deployment operation.

Post-war West Germany of course had to rebuild its transport requirements. It was forced initially to look to available foreign sources. The Noratlas fitted the bill, and 187 aircraft were ordered. Under the agreement, the first twenty-five were made in France and the remainder in Germany by Flugzeugbau Nord company under contract. The company had been involved in the Noratlas project in the design and manufacture of most of the fuselage. Finding early on that the aircraft Nord 2501D was not meeting requirements, Germany became the main source of used aircraft. Portugal was a main customer for the ex-German aircraft.

The Israel Air Force purchased three examples of the N-2501IS, apparently under pressure – if they did not purchase them, France would not allow them to purchase twelve of its Dassault Ouragan jet fighters. Israel was in the middle of the Suez Crisis in 1956 and needed to procure military aircraft. Many countries would not supply arms to Israel at that time, so it really had no choice regarding the Noratlas purchase. However, it soon came to realise how versatile the Noratlas was and purchased nineteen more, three N2501ISs and sixteen ex-German N2501Ds, primarily for cargo and paratroop transport. Some aircraft were converted for other uses, such as improvised bomber aircraft for long-range strike missions into Egypt and maritime reconnaissance during the Six Day War of 1967.

During 1970 the Hellenic (Greece) Air Force received fifty surplus aircraft as part of reparations from Germany. It was also part of a NATO military assistance programme. In 1974, the Noratlas fleet airlifted troops to the island of Cyprus to combat the Turkish invasion. This enabled Greece to retain control of the strategically important Nicosia International Airport.

The Hercules Powered Aircraft

The Portuguese Air Force operated a large number of Noratlas aircraft from 1960 on. Purchases were made from UAT, a civilian carrier, Nord company, and the German Air Force. They were used by Portuguese squadrons operating until 1977 in such areas as Portuguese Angola, Mozambique, and Guinea. This was all part of the extremely lengthy Portuguese Colonial War across the three African theatres. The Noratlas was used for tactical transport missions as well as delivering paratroopers for airborne assault. After independence, some Noratlas aircraft were given to the governments of these new countries.

In 1951 a civilian version was under development, known as the N-2502A/B, for some domestic and overseas customers.

Variants

Model	Aircraft Description
N2500	Prototype, Gnome et Rhône 14R (1,600 hp) engines, one built.
N2501	Production, French Air Force, SNECMA Hercules 739 1,521 kw (2,068 hp), five prototypes, 208 aircraft built.
Nord 2501A	Civil transport version, UTA, SNECMA 758/759 Hercules 1,230 kw (1,650 hp), four built, converted to N2502.
Nord 2501D	Production, German Air Force, German components, 186 built (25 French built, 161 German built).
Nord 2501E	Flight testing, two Turbomeca Marbore II turbojet engines, one built.
Nord 2501IS	Israeli Air Force, some Israeli components, 6 built.
N-2501 Gabriel	Armée de l'Air, SIGINT/electronic warfare platforms, 8 produced.
Nord 2501TC	Transvalair, 3 produced.
Nord 2501*	Air Algeria, UTA, Hercules 758/759 1,230 kw (1,650 hp),
Nord 2502A*	UTA five built, two conversions from 2501A.
Nord 2502B*	Air Algeria one built, two conversions from N2501A

The Hercules: The Other Engine That Helped Win the War

Model	Aircraft Description
Nord 2502C	Indian airline (intended), one prototype.
Nord 2502F	Portuguese Air Force, six conversions from Nord 2502
Nord 2503	Two 1,864 kw (2,500 hp) Pratt & Whitney R-2800-CB17 radial piston engines. One conversion from one of the Nord 2501 prototypes.
Nord 2504	French Navy, one built.
Nord 2505	French Navy, none built.
Nord 2506	STOL performance project, one conversion, one built.
Nord 2507	Search & Rescue, planning stage.
Nord 2508*	Modified Nord 2503, two prototypes, one conversion, one built.
Nord 2508B	Cargo transport version of the Nord 2508.
Nord 2509	Not built.
Nord 2510	French Navy, not built.
Nord 2520	Enlarged Nord 2502 with better cargo capacity, planning stage.

* with two Turbomeca Marbore IIE auxiliary turbojets.

Operators Military: Angola, Republic of the Congo, Djibouti, France, Germany, Greece, Israel, Mozambique, Niger, Nigeria, Portugal, Rwanda, Uganda.

Operators Civil: Algeria, Ecuador, France.

Nord 2501
Crew: Four or Five
Max takeoff weight: 20,603 kg (45,422 lb)
Fuel capacity: 5,090 L (1,120 imp gal)
Powerplant: Two SNECMA manufactured Bristol Hercules 738 1,558 kw (2,089 hp), Hercules 739 with torquemeters, Hercules 758/759 with reversible-pitch propellers
Propellers: 4-bladed constant-speed, fully feathering, non-reversible (Hercules 738/739)

Performance
- Maximum speed: 440 kmh (270 mph) at 21,000 kg (46,297 lb)
- Cruise speed: 335 kmh (208 mph) at 3,000 m (4,921 ft)
- Range: 2,500 km (1,600 miles)
- Service ceiling: 7,500 m (24,600 ft)

Northrop Gamma 2L

First Flight: 1932
Built by: Northrop, Production: 1, Total: 60

Introduction: 1932 The Northrop Gamma was a single-engine, all-metal monoplane cargo aircraft used in the 1930s. One, c/n 3471, was built and sold to the Bristol Aeroplane Company for testing the 962 kw (1,290 hp) Bristol Hercules Mk.1SM engine driving a three-blade propeller/flying test bed. Towards the end of its commercial life, it saw service as the A-17 light bomber.

Development: It was a direct descendant of the successful Northrop Alpha.

Design: It was a low-wing, fully enclosed cockpit aircraft which was fairly unique for its time. It had aerodynamics fairings, spats, on its fixed landing gear.

Operational history: Constructor number 3471 used as test bed for Hercules Mk ISM. Trans World Airlines used it as a mail plane, but its claim to fame was as a flying laboratory and record-breaking aircraft. It would eventually develop into the Northrop A-17 light attack aircraft and saw service with Spanish and Chinese Air Forces. In 1922, it set the record for Los Angeles to New York, and again in 1935, with the famous Howard Hughes in just over nine hours and twenty-six minutes. However, the most famous aircraft was *Polar Star*, which attempted to fly the first trans-Antarctic flight and just came up short by 40 km (25 miles) after running out of fuel. The aircraft rests in the Smithsonian National Air and Space Museum.

Variants: There were fourteen variants built, which performed a variety of roles from private (Texaco), Antarctic exploration as *Polar Star,* attack, cargo, flying laboratory, light bomber, two-seat race version, Sperry automatic test bed, two seat trainer with retractable landing gear, Bristol Hercules engine testing, Japanese study, Spanish coastal patrol, and aerial reconnaissance aircraft.

Operators Military: China, Spain, US.

Operators Civil: Manchukuo National Airways, Trans World Airlines.

Gamma 2D: statistics for 2D, powerplant for 2L
Crew: pilot
Max takeoff weight: 3,334 kg (7,350 lb)
Fuel capacity: unknown
Powerplant: One Bristol Hercules I 858 kw (1,150 hp)
Propellers: 2-bladed

Performance
- Maximum speed: 359 kmh (223 mph) at 1,900 m (6,300 ft)
- Cruise speed: 328 kmh (204 mph)
- Range: 3,170 km (1,970 miles)
- Service ceiling: 7,100 m (23,400 ft)

Northrop 8A-1

First Flight: 1935 (A-17)
Built by: Northrop Douglas, Production: 96, Total: 411
Operational: 1938–1944

The Hercules Powered Aircraft

The Northrop A-17 was a development of the Northrop Gamma 2F built by the Northrop Corporation for the United States Army Corps. It was called the Nomad when in service with the British Commonwealth during the Second World War.

Development: The A-17 was developed from the Gamma 2F, which itself had a revised tail and wing flaps, cockpit canopy, and semi-retractable landing gear from the Gamma 2C. The A-17 was further modified with perforated flaps, fixed landing gear with partial fairings, internal bomb bay, and external bomb racks. This in turn became the A-17A with fully retractable landing gear. The Douglas Aircraft Company took over the Northrop Corporation and the aircraft was renamed the Douglas Model 8.

Design: two seat, low-wing, monoplane.

Operational history: Among the military operators listed, Sweden purchased a licence for production of a Bristol Mercury-engined version. Hence, it was natural that Bristol would purchase a Swedish version, the Model 8A-1, to use as a test bed for its Bristol Hercules engine.

Variants: There were ten variant aircraft with designations of A-17, Model 8A, and Nomad.

Operators Military: Argentina, Canada, Republic of China, Iraq, Netherlands, Norway, Peru, South Africa, Sweden, US.

A-17A Statistics not for Bristol Hercules test aeroplane
Crew: (Two) pilot, gunner
Max takeoff weight: 3,328 kg (7,337 lb)
Fuel capacity: unknown
Powerplant: One Bristol Hercules for test purposes
Propellers: unknown

The Hercules: The Other Engine That Helped Win the War

Performance
- Maximum speed: 332 kmh (206 mph)
- Cruise speed: 274 kmh (170 mph)
- Range: 1,046 km (650 miles)
- Service ceiling: 5,915 m (19,400 ft)

Armament
Guns: Four .3 in (7.62 mm) fixed forward Browning machine guns, one .3 in (7.62 mm) trainable rear machine gun
Bombs: Internal plus external bomb racks 544 kg (1,200 lb) total

7

THE HERCULES IN ROYAL AIR FORCE SERVICE

The Bristol Hercules engine was of course designed for the Royal Air Force in the knowledge that Great Britain's fighting force may soon have to go to war, and there was a pressing need for powerful and reliable engines. The RAF was born of one world war and now looked like it would have to fight another one twenty-one years after the first. It was in these later stages of the inter-war period that the Bristol Hercules engine was coming of age. Would it be sufficiently reliable and powerful enough to contribute in a meaningful way to the war effort? Would the aircraft manufacturers consider it as a suitable power plant for their developing aircraft? Would the RAF make these aircraft part of their inventory to meet their operational requirements? What had the RAF learned from past acquisitions?

We must go back nearly 200 years to discover where the concept of aerial bombardment originated. It was 1760, and the Montgolfier brothers of France were experimenting with sending paper envelopes confining hot air into the skies. Twenty-three years later, the first human ascent in a balloon would occur over Paris. By the end of that century a tethered hydrogen balloon was being used to observe army movements during the French Revolution. In 1861, the Union Army Balloon Corps was

The Hercules: The Other Engine That Helped Win the War

observing the Confederate Army during the American Civil war. This worked well on calm days, but it would take an engine, and the addition of stabilisers, to make it more manoeuvrable on windy days, and not be restricted to the tied-down location. Just prior to the First World War, Italy and Libya were involved in skirmishes. It is thought that the first example of aerial bombardment were bombs dropped from an Italian dirigible on Libya during this conflict. General Giulio Douhet wrote a famous bombing treatise, *The Command of the Air*, following the Italian experience in Libya.

The air arm of the Royal Navy formed in 1910 under unusual circumstances. It was during the Mayfly rigid airship project, which was later abandoned, that the Royal Aero Club offered two aircraft and instructors at Eastchurch on the Isle of Sheppey for the four selected initial candidates. These officers had to be unmarried and able to pay the membership fees of the Royal Aero Club! The Air Battalion Royal Engineers (ABRE) was founded in 1911 and was comprised of two companies, the airship Company at Farnborough, Hampshire, and the aeroplane Company at Larkhill, Salisbury Plain, flying the Bristol Boxkite. This was the first connection of government aeroplane requirements with the Bristol and Colonial Aeroplane Company, later the Bristol Aeroplane Company. This working relationship was still in effect twenty-five years later when the Bristol Hercules engine appeared. The British Imperial Defence Staff, acknowledging the effective use of aircraft by Italy against the Ottoman Empire in Libya, recommended the establishment of a separate flying corps.

In 1911, the British Army and Royal Navy together could field three airships and half a dozen aircraft. France had over 200 aircraft and Germany had thirty airships. The Royal Flying Corps (RFC) was established on 13 April 1912 by royal warrant. It had two Wings, the Military Wing, the army element established 13 May 1912, and the Naval Wing. The RFC was also comprised of a Central Flying School, a Reserve, and the Royal Aircraft Factory to manufacture its military aircraft. The RFC was tasked with fighter (scout), bomber, and observation duties. The RFC Squadrons played a vital role in battlefield management, being the eyes of the army. The aircraft, organised in squadrons, were assigned communication, reconnaissance, observation (aerial photography), and artillery direction duties.

The RFC looked good on paper, but the history of the army and navy rivalry would prevent full cooperation and amalgamation. The Admiralty would never allow it; after all, it was the senior service, control of its aviation affairs could not be handled by any other organisation. In fact, the Royal Flying Corps, Naval Wing existed only on paper. The Navy itself trained its own pilots and, ignoring the Royal Aircraft Factory at Farnborough, ordered its own aircraft directly from the manufacturer. Lack of co-ordination of effort by naval and military aviation caused the necessary establishment of an Air Committee in 1912, immediately after the formation of the Royal Flying Corps. This committee was comprised of members of the two war ministries and was limited to recommendations to the Admiralty Board and Imperial General Staff, which had to ratify them before presenting them to the War Office. In 1913, the balloons were transferred to the Royal Navy.

Two years later, the situation became official; The Royal Naval Air Service (RNAS) was founded on 1 July 1914. The position at the beginning of the First World War was that the RFC was in effect part of the army, with RNAS operating separately. The RFC was not an independent arm as intended. The RFC and the RNAS were at odds, which was affecting the design and supply of aircraft and the defence of Britain. The RNAS immediately set up a chain of ten naval air stations around the coastline to be operational centres for airships and aircraft. With a lack of armed fighter type aircraft, the defence was largely ineffectual against the Zeppelins in 1915. Initially, the RNAS aircraft development was unsurprisingly for operations over water, which in due course led to ship-launched and crane-recovered aircraft, totally dependent on sea conditions.

The Hercules: The Other Engine That Helped Win the War

Both the RFC and RNAS went to France with the British Expeditionary Force and operated separately; the RNAS in a more aggressive, individual, and psychological bombing role and the RFC in the army support reconnaissance role, using the BE2 built at the Royal Aircraft Factory. By 1915, the RFC now had three Wings of eight squadrons each in France. The armed scout made an appearance, and the RFC was equipping with independent manufacturers' aircraft, such as Vickers, Martinsyde, and Bristol. Colonel Hugh Trenchard took over the RFC and changed the focus of the aircraft from the passive role to the armed scout, the Royal Aircraft Factory FE2, and bomb dropping operations.

An ineffective committee during the first years of the First World War resulted in the formation of an equally ineffective Joint War Air Committee in 1916. The Air Board in 1917, driven by higher profile member Lord Cowdray, was slightly more successful and resulted in specifications for the Airco DH.10 day bomber and the Fairey III and Short N.2B floatplane bombers. By 1917, the air superiority was changing in Britain's favour with such fighting aircraft as the Sopwith Pup and Sopwith Camel. This was joined by the De Havilland DH4 and Handley Page O/400 bombers to take the fight to the enemy. It should be noted that the RFC was also involved in overseas campaigns in Italy and Palestine. The RNAS were using large American-built flying boats for sea operations.

But 1917 was not all smooth going; Zeppelins accompanied by twin-engined Gotha bombers were now causing damage with day and night raids. General Smuts, of Boer War fame, now a loyal subject of the British Empire, was appointed to look into the situation and to see what could be done in response. His recommendation was, once again, to form one organisation to look after the air requirements of Great Britain.

The Air Force (Constitution) Bill was passed by Parliament on 29 November 1917 because of his report. The first Air Council was formed on 3 January 1918 and very shortly faced resignations. Feelings were running very high regarding how this new organisation was going to function, especially now that the Germans were focusing their ground attack on Amiens and the Allies were waiting for the American troops to arrive. The RFC concentrated their attacks on small areas rather than the whole front, and that approach seemed to be more effective. The German Army then concentrated their attention to the North instead, and it was at this time that the Royal Air Force was formed on 1 April 1918. The German effort had run out of resources and the plans stalled. By November 1918 when the First World War ended, the RAF was seven months old. The Royal Flying Corps and Royal Naval Air Service individually had expanded rapidly under the War Office and Admiralty to 103 airships and to over 22,000 aircraft in the four years of hostilities, phenomenal growth, what an effort by so many, many people.

Part of that expansion was due to the Bristol and Colonial Aeroplane Company's contribution. It started four years previously when an aircraft design by Henri Farman was published, with dimensions, in the aeronautical press. The Boxkite was a successful commercial venture but had limited future potential.

Bristol Aircraft WW1

Year	Aeroplane	Type
1910	Bristol Biplane (Boxkite)	Pusher type trainer
1911	Bristol Monoplane	Experimental monoplane, two built
1911	Bristol Biplane Type T	Pusher type cross-country racer, 5 built
1912	Bristol Side by Side Monoplane	Trainer, side by side cockpit, 6 built
1913	Bristol-Coanda Two-Seat Biplane	Long range aircraft (5 hr), advanced trainer
1914	Bristol Scout Types A, B, C, D, SSA, S2A	Racing, fast reconnaissance, fighter aircraft, 374 built

The Hercules: The Other Engine That Helped Win the War

Year	Aeroplane	Type
1916	Bristol M1 Monoplane Scout	Only British monoplane fighter to reach production during the War, 130 built
1916	Bristol F2A, F2B	Reconnaissance aircraft that became a successful fighter aircraft, 5,329 built
1917	Bristol MR1 Metal Biplane	Experimental, 2 built
1918	Bristol Scout E, F	Roy Fedden of Cosmos Engineering tested his Cosmos Mercury engine in an F1, 4 built
1918	Bristol Braemar, Pullman, Tramp	Heavy bomber, steam transport (Tramp) 5 built

The bombing started by the RNAS would be continued and enhanced by the RAF, which had Major General Sir Hugh Trenchard as its first leader. Trenchard was in charge of the Independent Bombing Force, former VIIIth Brigade, which contributed to the final victory, but more significantly, important lessons were learned for future aerial warfare.

By 1919, the disbanding of squadrons had begun, with the Admiralty still smarting after the loss of control of their air services. A political change to ministers resulted in Winston Churchill in charge of the War Office and Air Ministry working with General Trenchard's RAF. Trenchard had his work cut out for him to preserve the junior service from takeover by the Navy and Army air services. He immediately set about establishing its own independent culture by establishing training facilities at Cranwell, RAF Cadet College, Lincolnshire, and a technical training facility at Halton, Cheshire. He also inaugurated a territorial arm, which would become the Auxiliary Air Force. This far-sighted foundation established by Trenchard firstly preserved the RAF, in its own right, and secondly allowed it to prepare for whatever eventualities lay ahead. Not one to forget those affected by the misfortunes of war, Trenchard established the RAF Benevolent Fund. The package was complete by the end of 1919.

The British Empire, although under pressure, still existed and the RAF maintained a presence in such areas as the North West Frontier of India, Iraq, Mesopotamia, Afghanistan, and Somaliland. This was an expensive but necessary operation for the British Empire to 'show the flag'. On the home front, his master stroke was to get the public on his side by making the RAF highly visible at the RAF Tournament in Hendon, London. Displays took place from 1920 to 1937 with up to 60,000 paying spectators at each event. It also instilled pride in the members of the RAF. This is not to say that the naval environment was being overshadowed. HMS *Argus* was a flat deck ship that had the capability of launching and recovering aircraft, the first aircraft carrier. Further RAF development resulted in the Fleet Air Arm being established 1 April 1924, for those RAF units that normally embarked on aircraft carriers. The Home Defence organisation consisted of the Special Reserve Squadrons and the Auxiliary Air Force.

In 1925, Trenchard established the University Air Squadrons to attract high calibre individuals to the Air Force rather than the other services by offering them the opportunity to fly while attending higher education. It goes without saying that the two renowned universities, Oxford and Cambridge, would each have a squadron. Training emphasis had shifted from 'how to stay out of trouble' to 'here is what you can do with this aircraft,' thanks to a former scout squadron commander, Robert Smith-Barry. Thanks to the organisation and the slightly overbearing presence of General 'Boom' Trenchard, the RAF was here to stay. In 1927, the Bristol Aeroplane Company produced the Bulldog. It was the most successful aircraft for the RAF during the inter-war period, powered by the Bristol Jupiter, Mercury, Perseus, and Aquila engines; 443 were built.

It should be mentioned that the Fleet Air Arm continued to develop under the auspices of the RAF by building six aircraft carriers in the 1920s. Many of the larger cruisers and battleships had catapult-launched floatplanes, with crane recovery on some ships possible. It did suffer from not keeping up with the advances of other nations, as the RAF focus was on their own bombers and fighters. A serious situation which came to light when the U-Boats appeared in 1939.

RAF prowess would be demonstrated to the public by such things as a formation flight from Cairo, Egypt, to Kano, Nigeria, and a flight in March 1926, from Cairo, Egypt, to Cape Town, South Africa. Schneider Trophy flights were undertaken by RAF pilots belonging to the High Speed Flight especially formed for the Trophy races. Another aspiration of the RAF was to set a long-distance record. It did so in 1933 with a non-stop flight from Cranwell, Lincolnshire, to Walvis Bay, South Africa, a distance of 8,544 km (5,309 miles) in fifty-seven hours and twenty-five minutes.

The end of the decade closed with the retirement of the 'Father of the RAF', General Trenchard. He had accomplished what he had set out to do: to establish an independent air force which could stand alone, with the desire 'to get the job done'.

Hard to believe today, but the post-war threat was deemed to be the traditional enemy, France. So, the focus of defence was across the Channel to Britain's closest neighbour. But the 1930s gave rise to two different sources of possible future threat, the totalitarianism of Mussolini in Italy, and Germany quietly rebuilding its military. The RAF had gone into a state of present strength existence, with little or no realistic increase in capability. The public were still recovering from the First World War. It had been a huge drain on the economy and a generation of young men.

In 1930, the British manufacturers were still producing biplanes that would fly in good weather clear of clouds to bring the fight to the enemy or defend home territory. The open cockpit was still in vogue, and the development of blind flying instruments was still in its infancy. The 'seat of the pants' flying by feel was coming to an end. The training would have to start again to prepare new pilots to fly by their instruments and not by feel. The Avro 504 was equipped with a blind-flying hood to train the pilots for the new instrumented aircraft.

The Hercules in Royal Air Force Service

Under the guise of the German airline Lufthansa, a secret plan was discovered to build German air forces. The RAF still had biplanes as this new threat emerged. Fortunately, the British aircraft manufacturers not only worked to government specifications but did some experimental work on their own. The Hawker and Supermarine aircraft were experimenting with monoplane fighters. At the same time, the Rolls-Royce company was learning from previous engines and had a very promising design in the works, the famous Merlin. It would go on to power many of the RAF's aircraft such as the well-known Hawker Hurricane, the Supermarine Spitfire, and the Avro Lancaster. A vital move occurred in the middle 1930s to make the RAF more cohesive and effective; the RAF established commands, namely Bomber, Fighter, and Coastal, among others, that would have connections to the Hercules engine. Later, during wartime in the 1940s, Transport Command would follow.

This is the same time period that Roy Fedden and the Bristol Aeroplane Company were working on the new concept sleeve-valve engines. The Bristol Perseus, 1932, and the Bristol Aquila, 1934, were leading the way for future sleeve-valve engines. They would be used by the RAF and Fleet Air Arm aircraft.

Bristol Perseus Sleeve-Valve Engine (1932)	
Manufacturer	Aircraft
Blackburn	Botha (RAF), Roc and Skua, Fleet Air Arm*
Bristol	Bulldog (RAF), Type 148 (Prototype)
De Havilland	Flamingo, Hertfordshire (Transport/General Communication)
Gloster	Goring (Prototype)
Hawker	Hart (RAF)
Saro	A33 (Prototype)
Short	Empire, Scylla (Airliners)
Victor	Velox (Airliner), Vildebeest (RAF light/torpedo bomber)
Westland	Lysander (RAF) (Army co-operation)

Bristol Aquila Sleeve-Valve Engine (1934)	
Manufacturer	Aircraft
Bristol	Bulldog (RAF), Bullpup (Prototype), Type 143 (Prototype)
Vickers	Venom (Prototype)

The fourteen cylinder, two-row sleeve-valve radial Bristol Hercules (1936) engine would be used by the RAF in a variety of aircraft. The Hercules would indeed be ready for the conflict and would certainly help to win the Second World War. It would eventually serve in all four commands, during the war years and for many years after. The Bristol Aeroplane Company had not only contributed to British aviation and the RAF by manufacturing aircraft but continued to develop engines in its engine division to supply the same market. The Bristol Mercury powered Bristol Blenheim (1936), a mid-wing, medium bomber, contributed to the success of the RAF pre-war buildup and the initial years of the Second World War; 6,215 were built. The following variety of Hercules powered aircraft would be assigned squadrons in one of four commands depending on the aircraft's role.

THE HERCULES IN RAF COMMANDS

Fighter Command
1 May 1936–17 November 1943
Motto: *Offence Defence*
HQ: RAF Bentley Prior, London
Fighter Command originated in the Fighting Area Group, part of the air defence of Great Britain. Fighting Area gained command status in 1932, and on 1 May 1936 was named Fighter Command. To enhance the interception role, No 60 Group RAF was established to control the radar detection and tracking units of the Chain Home network. It was now time to replace the obsolete Bristol Bulldog, Gloster Gauntlet, and Hawker Fury aircraft. Bristol engines were associated in different capacities

with all three aircraft. These aircraft were replaced by the famous Hawker Hurricane and Supermarine Spitfire monoplane fighters.

Fighter Command's 'Finest Hour' was during the German Luftwaffe offensive, Operation *Sea Lion*, in the summer of 1940, which was the prelude to a seaborne invasion. Enough has been written about The Battle of Britain and suffice to say here that Luftwaffe air superiority never occurred, and the invasion plans were abandoned. Fighter Command then planned short penetration flights over the English Channel and Northwestern France to keep the Luftwaffe occupied.

The 'Blitz' by German bombers continued and this is when the Bristol Hercules powered Bristol Beaufighter IF came into prominence as a night fighter with its own airborne interception radar, in co-operation with Ground-Controlled Interception (GCI). The first kill by a radar-equipped Beaufighter night-fighter was of a Junkers Ju 88 on the night of 19/20 November 1940. By 1941, with improved radar capability, it was responsible for over half of the enemy aircraft destroyed. In May 1941, the Luftwaffe transferred its attention to the Eastern Russia Front after failing to secure the skies over Britain. The Beaufighter saw extensive service during the war with the RAF, fifty-nine squadrons, and Fleet Air Arm, fifteen squadrons. For the first two years of hostilities the Bristol Beaufighter was an effective combination of the Bristol Hercules engine and the Beaufighter airframe, a winning combo for Bristol.

Coastal Command
14 July 1936–27 November 1969
Motto: *Constant Endeavour*
HQ: Northwood, London
Coastal Command originates with the RAF Coastal Area, which was founded in 1919 and responsible for anti-submarine warfare (ASW), commerce raiding, aerial reconnaissance, search and rescue, and weather reconnaissance. In conjunction with the 1914 RNAS, the Admiralty introduced new types of aircraft and even encouraged competition and innovation among

The Hercules: The Other Engine That Helped Win the War

manufacturers within the Admiralty's parameters. There was a debate as to whether the smaller and cheaper floatplanes or the larger, expensive, but longer range, flying boats were the best solution. Maritime warfare's emphasis shifted from the large flotillas of large ships to the challenge of the U-boats and how to find them and, if possible, destroy them.

Maritime air operations became the responsibility of the RAF upon its foundation in April 1918. The inter-war struggle between the Admiralty and RAF would escalate during the 1920s and 1930s as the emphasis shifted to land-based aircraft for attack, the bombers, and for defence, the fighters. In 1936, the 'Area' was raised to command level, and Coastal Command was created. Its mission was the protection of Allied convoys from the prowling U-boats and to sink the enemy's shipping. The following RAF aircraft used the Bristol Hercules engine to carry out that mission.

Coastal Command would use the Beaufighter IC successfully with Nos 252 and 272 Squadrons RAF based in Malta for Mediterranean operations. The VIC carried a 450 mm (18 in) torpedo externally. The first successful torpedo attack was by No 254 Squadron RAF, RNAS Prawle Point, Devon, sinking two ships off Norway. The main production mark of the Beaufighter was the TF Mark X with Bristol Hercules Mark XVII engines. This variant was commonly known as the 'Torbeau'. These air-to-surface radar-equipped aircraft could also be equipped with RP-3 27 kg (60 lb) rockets. The Strike Wing of Coastal Command based at RAF North Coates, Lincolnshire, was comprised of Nos 143, 236, and 254 Squadrons RAF. They developed tactics that used large formations of Beaufighters to suppress flak while the 'Torbeaus' attacked low level with torpedoes. This Strike Group was the largest anti-shipping force of the Second World War and sank half the tonnage of all Strike Groups for the loss of 120 Beaufighters and 241 aircrew killed or missing.

For a discussion of the Handley Page Halifax, see entry under Bomber Command.

The Saro Lerwick armed with bombs/depth charges and three-gun turrets, was a medium range flying boat for ASW, convoy escort (CE), and reconnaissance (R). It was difficult to fly as it was unstable and had vicious stall characteristics; only twenty-one were built. The Lerwick Mk 1 served from December 1939 until April 1941 with No 209 Squadron RAF in Oban, Western Scotland, patrolling the Atlantic Ocean. By the end of 1942 it was declared obsolete.

The Short S26 was a large transport flying boat designed for transatlantic civilian flight that was diverted to RAF use due to the outbreak of hostilities. By late 1941, the conversion by Blackburn Aircraft Limited, Dumbarton, Scotland, to military use was complete and included the addition of radar. The S26, sometimes referred to as 'G-Boats', served with No 119 Squadron RAF from March until October 1941 based at RAF Bowmore, Argyll and Bute, Strathclyde Region, Scotland. Bowmore is situated on the island of Islay, which gave sheltered access to the Atlantic for patrols. Three aircraft were built.

The Short Seaford was a long-range maritime patrol bomber developed from the Short Sunderland specifically for service in the Pacific Ocean. Aerodynamic stability problems caused extensive changes to the Sutherland and hence the new name. Eight were built too late for wartime operations, but six were delivered to No 201 Squadron RAF at RAF Pembroke Dock, Pembrokeshire, Wales, in 1946, for a short operational trial period.

Bomber Command
14 July 1936–30 April 1968
Motto: *Strike Hard Strike Sure*
HQ: 1936–40 RAF Uxbridge, 1940–68 RAF High Wycombe
The founding of Bomber Command coincided with the first running of the Hercules engine. This is the command where the Hercules would contribute the most to the war effort. In the first few weeks of the First World War the German Naval Airship Department's Zeppelins dropped bombs on the citizens of Liège and Antwerp, Belgium. This was followed by aerial bombardment of England for the next three years, with the airships being complemented by Gotha 'G' four engine bombers in the latter

The Hercules: The Other Engine That Helped Win the War

years of the conflict. The precedence for aerial bombardment had been established not only by airships but also by reconnaissance and bomber aircraft used by both sides in the conflict.

In 1936 the Royal Air Force created Bomber Command to build up an organisation of personnel and equipment to meet 'possible' future requirements. The politicians and purveyors of the British Government purse still believed that diplomacy, in spite of all contra-indications, could avert the looming crisis.

Bomber Command in the Second World War was a direct descendant of the First World War strategic bombing forces. The strategic bombing philosophy was in the hands of Bomber Command, subject to the overall aims of the war effort. Initially, as the threat grew it was the junior partner in the build- up of arms. It was the navy that enjoyed priority, followed by the defensive air force. The striking power of Bomber Command in the early years was limited to leaflet dropping, defensive patrols, and a disastrous attempt at daylight raids on the heavily protected German Navy. Bomber Command initially lagged behind in manpower and advanced aircraft but continued to grow, and by 1942, under a new Commander-in-Chief Air Marshal A. T. Harris, the tables turned with night area bombing done by heavier and more numerous bombers.

Bomber Command eventually increased to thirteen Groups, each with its own headquarters under the command of at least a Group Captain, an Air Vice-Marshal if a front-line Group. No. 2 Group was formed in March 1936, a scant three years before its first operational mission in September 1939. Perhaps the most famous of all the Groups was No. 8 Group, the Pathfinder Force. It was in these groups that the following Hercules powered aircraft contributed the most to the war effort.

The Avro Lancaster BII entered service in December 1942 with No 44 (Rhodesia) Squadron RAF at RAF Waddington, Lincolnshire, and shortly thereafter performed its first operational trip on 2 March 1942. It was to deploy naval mines in the vicinity of Heligoland Bight. The first bombing raid was on 10 March 1942 to the city of Essen, North-Rhine-Westphalia. The 7,377 Lancasters

The Hercules in Royal Air Force Service

delivered a total of 552,124 tonnes (608,612 tons) of bombs during 156,192 sorties. These missions were flown by sixty-six squadrons. In addition, there were sixteen Heavy Conversion Units, three Lancaster Finishing Schools, one Operational Training Unit, and two conversion units. It would carry the 12,000 lb *Tallboy* and 22,000 lb *Grand Slam* bombs. The Hercules VI or XVI powered the BII variant. The Hercules engine was fitted to the Lancaster due to the concern that there would not be enough Rolls-Royce Merlin engines to sustain Lancaster production and that the American-built Packard engines were susceptible to the U-Boat activity against the North Atlantic convoys bringing the engines to Britain.

The Hercules engine version was built by Armstrong Whitworth in their factory at Whitley, Coventry, West Midlands. It entered service with No 61 Squadron RAF in October 1942 for initial service tests, which revealed an important limitation. During a mixed aircraft Lancaster raid in January 1943 the force of BIs was at 6,706 m (22,000 ft) but the BIIs could only climb to 5,486 m (18, 000 ft). Just 300 of the BII were built, less than 4% of the total, due to this performance limitation. Issued to No 115 Squadron RAF, the Lancaster BII was a replacement for the Vickers Wellington bombers. As noted by the dates below, the BII remained in squadron service a relatively short time before being replaced by the better performing BI or BIII Rolls-Royce Merlin versions. See the Halifax below for a reverse of that sequence.

The Lancaster BII was in service with three RAF squadrons, Nos 61, 115, and 514. No 61 Squadron RAF was at RAF Syerston, Nottinghamshire, flying the BII from October 1942 until March 1943. No 115 Squadron RAF was at RAF East Wretham, Norfolk, RAF Little Snoring, Norfolk, and RAF Witchford, Cambridgeshire, flying the BII from March 1943 until May 1944. No 514 RAF Squadron was at RAF Foulsham, Norfolk, and RAF Waterbeach, Cambridgeshire, flying the BII from September 1943 until July 1944. In addition, the BII was assigned to Nos 408 (Goose), 426 (Thunderbird), and 432 (Leaside) Squadrons RCAF, all based in Yorkshire.

The Hercules: The Other Engine That Helped Win the War

The Handley Page Halifax was a four-engine heavy bomber developed in response to the same specification request as the two-engine Avro Manchester. The first flight was in October 1939 and the first of 6,176 aircraft entered RAF service on 13 November 1940. The first squadron to have the Halifax was No 35 Squadron at RAF Linton-on-Ouse, North Yorkshire. The first of the 82,773 WWII operations, dropping a total of 203,397 tonnes (224,207 tons) of bombs, was a night raid on the docks at Le Havre, Normandy. By 1942, No 35 Squadron RAF was part of the Pathfinder Force. In all, the Halifax saw service in forty-two RAF squadrons. No 4 Group, Bomber Command, operated the Halifax and would continue to do so until the end of hostilities. At its peak the Halifax was operated by seventy-six Bomber Command squadrons. The Halifax would see support duties as a glider tug, electronic warfare for No 100 Group RAF, and Special Operations Executive parachuting arms and agents into enemy-held territory. It saw service with Coastal Command in anti-submarine warfare, reconnaissance, and meteorological operations.

The Halifax Mk III and IV were Hercules engine versions; 2,091 and 643 being built respectively. An example of how unpredictable aircraft design could be during this period was the case of the Halifax Mk III. This saw the Halifax switch from the Rolls-Royce Merlin to the Hercules engine. The performance of the Halifax was improved by the swap to the Hercules in much the same way as the Lancaster had been improved by the change to the Rolls-Royce Merlin! The Halifax LW682 shot down near Geraardsbergen, Belgium, now makes up part of the roof of the Bomber Command Memorial in London. This is thanks to Halifax 57 Rescue of Canada, who in 1997 melted the airframe for the material. This same group is restoring a Halifax to ground running condition, see Chapter 8.

The Short Stirling was a four-engine heavy bomber designed from the start to have four engines, unlike the Avro Lancaster and Handley Page Halifax, which started life on the drawing board as two-engine aircraft. The Stirling was bigger than both these aircraft and first flew in May 1939. The first was delivered to No 7 Squadron at RAF Leeming, North Yorkshire, and was operational by January 1941. The following month, three Stirlings flew a mission against the fuel dumps at Vlaardingen, near Rotterdam, Netherlands. By the end of the year, 150 Stirlings were in service with the RAF. No 7 Squadron also played a pioneering part in the formation of the Pathfinder Force, the target finding and marking force. Stirlings were part of No 161 Squadron RAF, the highly secretive unit that was tasked with missions for the Special Operations Executive and Secret Intelligence Service. It would serve with twenty-eight RAF squadrons.

The Vickers Wellesley was a medium bomber introduced in 1937 and commenced service with No 76 Squadron RAF Finningley, South Yorkshire. By 1939, it had been phased out on the Home Front but saw continuing service in the East African Campaign, based at one time in Egypt and Sudan. No 47 Squadron RAF operated the Wellesley until 1942 on maritime reconnaissance over the Red Sea.

The Vickers Wellington was introduced to RAF service in 1938, serving with Nos 9, 37, 38, 99, 115, and 119 Squadrons RAF as part of No 3 Group, Bomber Command. Less than twenty-four hours after the outbreak of war, Wellingtons of Nos 9 and 149 Squadrons RAF performed the first bombing raid of the war, on shipping at Brunsbuttel, Schleswig-Holstein, Northern Germany. During December, bombing raids by No 99 Squadron RAF concentrated on the German naval bases at Schillig Roads and Wilhelmshaven. Otherwise known as the 'Wimpy', it was the first RAF bomber to drop the 4000 lb 'Blockbuster' bomb. The Wellington also provided service to Coastal Command during the Battle of the Atlantic. Superseded in the European Theatre, it continued to serve the RAF in the Middle and Far East.

Transport Command
25 March 1943–01 August 1967
Motto: *I Strike by Carrying*
HQ: RAF Upavon, Wiltshire

The Hercules: The Other Engine That Helped Win the War

Transport Command had its direct origins in Ferry Command, which was the secretive command established to ferry aircraft from their manufacturers in Canada and the United States to the frontline units. The task was to get foreign aircraft operational as quickly as possible and minimise losses. It was taking too much time to disassemble the aircraft, load them on ships which were subject to attacks by U-Boats, offload them, assemble and test fly them, and when ready for service deliver them to their squadrons. Lord Beaverbrook, a Canadian and Minister of Aircraft Production, quietly arranged for civilian crews to ferry the aircraft across the Atlantic. Initially working with the Canadian Pacific Railway Corporation, it became the Atlantic Ferry Organisation, 'Atfero'. It was raised to command status and became Ferry Command on 20 July 1941 and succeeded in ferrying more than 9,000 aircraft across the Atlantic. On 25 March 1943, it was absorbed into the new Transport Command as No 45 (Atlantic Ferry) Group. Other groups were responsible for movements associated with Egypt, India, Gold Coast, and arrivals in Britain.

The Armstrong Whitworth Albemarle was originally designed as a medium bomber, but with superior bombers already operational, its role was changed to reconnaissance and transport duties. This redesign caused production delays and it was not until January 1943 that No 295 Squadron RAF at RAF Harwell, Oxfordshire, received the aircraft in quantity. No 296 Squadron RAF at RAF Hurn, Dorset, was tasked with dropping leaflets over France in 1943. No 570 Squadron RAF at RAF Harwell, Berkshire, towed the Airspeed Horsa gliders during the Normandy invasion on 6 June 1944 along with Nos 295 and 296 Squadrons RAF. The Albemarle served with three Heavy Glider Conversion Units, two Ferry Training Units, and in various trials by aircraft companies.

The Avro York was a transport aircraft developed using parts of the Avro Lancaster. Only one was built using the Hercules engine as a prototype. The aircraft was used during the Berlin Airlift and for VIP transport. 258 aircraft were built serving eleven RAF squadrons and ten sundry units.

The Handley Page Hastings was a troop carrier and freight transport aircraft built specifically for the RAF and when introduced was the largest transport aircraft ever designed for the service. Thirty-two of the 151 built took part in the Berlin Airlift, working alongside the Avro York, which it eventually replaced. It took part in the Suez Canal crisis, Nos 70, 99, and 511 Squadrons RAF dropping paratroopers and supplies in the Indonesian conflict, as well as working as a much-needed long range transport. No 47 Squadron based at RAF Dishforth, North Yorkshire, was the first squadron to receive the Hastings. Four Hastings were assigned to No 24 Squadron RAF at Brize Norton, Oxfordshire, as VIP transports. In 1950, the weather reconnaissance version was based at RAF Aldergrove, Northern Ireland, with No 202 Squadron RAF, until replaced by satellites in 1964.

The Vickers Valetta was a twin-engine, tailwheel military transport which rapidly replaced the Douglas Dakota both at home in England and abroad in the Middle East and Far East. It had VIP and Navigation Training versions. Eighteen squadrons used the Valetta including No 30 Squadron RAF based at RAF Oakington, Cambridgeshire. The Valetta was used in crisis areas such as Egypt, Aden, and the Malayan conflict to drop paratroopers and supplies. 262 aircraft were built.

The Vickers Varsity was a twin-engine crew trainer operated by the RAF from 1951 until 1976. Based on the Viking and Valetta, it had tricycle landing gear. It was introduced to replace the Vickers T10 Wellington trainer at RAF Swinderby, Lincolnshire, to train pilots on multi-engine aircraft. It was operated by No 201 Advanced Flying School RAF. It also saw service as a navigation trainer with the Central Navigation and Control School and later with the Bomber Command Bombing School to train the V-bomber crews for the new nuclear bomb capability.

The Hercules was part of the RAF's history from just before the Second World War, helping to win the war in many of its Command's aircraft, and transitioning Transport Command post-war into the emerging jet age. The engine certainly served with distinction in the RAF over thirty years.

The Hercules: The Other Engine That Helped Win the War

The following table indicates some of the Royal Air Force Second World War bomber squadrons that operated Bristol Hercules powered aircraft. These aircraft also saw exemplary service with other Allied air forces. This war effort, of course, would include the 'erks' to service the engine, the equipment and facilities (spare parts) for engine support, and the all-important supply lines to the intentionally scattered stations. It really was 'the other engine that helped win the War.'

Squadron	Aircraft	Date
7	Short Stirling I	Aug. 1940–Jul. 1943
	Short Stirling III	Mar. 1943–Jul. 1943
10	Handley Page Halifax BIII	Mar. 1944–Aug. 1945
12	Vickers Wellington III	Aug. 1942–Nov. 1942
15	Vickers Wellington IC	Nov. 1940–May 1941
	Short Stirling I	Apr. 1941–Jan. 1943
	Short Stirling III	Jan. 1943–Dec. 1943
35	Handley Page Halifax BIII	Dec. 1943–Mar. 1944
37	Vickers Wellington III	Mar. 1943–Apr. 1943
38	Vickers Wellington VIII	Mar. 1942–Oct. 1943
	Vickers Wellington X	May 1942–Oct. 1943
	Vickers Wellington XI	Jul. 1943–May 1944
	Vickers Wellington XIII	Oct. 1943–Jan. 1945
	Vickers Wellington XIV	Jan. 1945–Jun. 1946
40	Vickers Wellington III	May 1942–Jul. 1943
	Vickers Wellington X	May 1943–Mar. 1945
70	Vickers Wellington III	Jan. 1943–Dec. 1943
	Vickers Wellington X	Jun. 1943–Feb. 1945
75	Vickers Wellington III	Jan. 1942–Oct. 1942
76	Handley Page Halifax B.III	Jan. 1944–Apr. 1945
	Handley Page Halifax B.VI	Mar. 1945–Aug. 1945
77	Handley Page Halifax B.III	May 1944–Mar. 1945

The Hercules in Royal Air Force Service

Squadron	Aircraft	Date
	Handley Page Halifax B.VI	Mar. 1945–Aug. 1945
78	Handley Page Halifax B. Mk III	Jan. 1944–Apr. 1945
	Handley Page Halifax B.VI	Apr. 1945–Jul. 1945
90	Short Stirling I	Dec. 1942–May 1943
	Short Stirling III	Feb. 1943–Jun. 1944
96	Handley Page Halifax C. Mk III	Dec. 1944–Apr. 1945
99	Vickers Wellington III	Oct. 1942–Nov. 1943
	Vickers Wellington X	Jun. 1943–Sept. 1944
	Vickers Wellington XI	Oct. 1943–Dec. 1943
101	Vickers Wellington III	Feb. 1942–Oct. 1942
104	Vickers Wellington X	Jul. 1943–Mar. 1945
115	Vickers Wellington III	Nov. 1941–Mar. 1943
142	Vickers Wellington IV	Oct. 1941–Oct. 1942
	Vickers Wellington III	Sept. 1942–Oct. 1943
	Vickers Wellington X	Aug. 1943–Oct. 1944
144	Bristol Beaufighter VI	Jan. 1943–May 1943
	Bristol Beaufighter X	May 1943–May 1945
148	Short Stirling IV	Nov. 1944–Dec. 1944
149	Short Stirling III	Feb. 1943–Sept. 1944
156	Vickers Wellington III	Mar. 1942–Jan. 1943
158	Handley Page Halifax B. Mk III	Jan. 1944–Jun. 1945
	Handley Page Halifax B. Mk VI	Apr. 1945–Jul. 1945
161	Armstrong Whitworth Albemarle I	Oct. 1942–Apr. 1943
166	Vickers Wellington III	Jan. 1943–Apr. 1943
	Vickers Wellington X	Feb. 1943–Sept. 1943
196	Vickers Wellington X	Dec. 1942–Jul. 1943
	Short Stirling III	Jul. 1943–Feb. 1944
	Short Stirling IV	Feb. 1944–Mar. 1946
	Short Stirling V	Jan. 1946–Mar. 1946
199	Vickers Wellington III	Nov. 1942–May 1943

The Hercules: The Other Engine That Helped Win the War

Squadron	Aircraft	Date
	Vickers Wellington X	Mar. 1943–Jul. 1943
	Short Stirling III	Jul. 1943–Mar. 1945
	Handley Page Halifax III	Feb. 1945–Jul. 1945
214	Vickers Wellington II	Jun. 1941–Jan. 1942
	Short Stirling I, III	Apr. 1942–Jan. 1944
215	Vickers Wellington X	Sept. 1943–Aug. 1944
218	Short Stirling I	Jan. 1942–Apr. 1943
	Short Stirling III	Apr. 1943–Aug. 1944
271	Armstrong Whitworth Albemarle I	Oct. 1942–Apr. 1943
295	Armstrong Whitworth Albemarle I	Oct. 1942–Apr. 1943
296	Armstrong Whitworth Albemarle I	Jan. 1943–Oct. 1944
297	Armstrong Whitworth Albemarle I	Jul. 1943–Dec. 1944
300	Vickers Wellington III	Jan. 1943–Apr. 1943
	Vickers Wellington X	Mar. 1943–Apr. 1944
301	Vickers Wellington IV	Aug. 1941–Apr. 1943
304	Vickers Wellington III, X	Jun. 1943–Jul. 1943
	Vickers Wellington XIII	Jul. 1943–Sept. 1943
	Vickers Wellington XIV	Sept. 1943–Jan. 1946
	Handley Page Halifax C.8	May 1946–Dec. 1946
305	Vickers Wellington X	May 1946–Sept. 1943
347	Handley Page Halifax B. Mk III	Jul. 1944–Apr. 1945
	Handley Page Halifax B. Mk VI	Mar. 1945–Nov. 1945
511	Armstrong Whitworth Albemarle I	Nov. 1942–Mar. 1944
513	Short Stirling III	Oct. 1943–Nov. 1943
514	Avro Lancaster II	Sept. 1943–Jun. 1944
517	Handley Page Halifax Met. Mk III	Mar. 1945–Jun. 1946
518	Handley Page Halifax Met. Mk III	Mar. 1945–Oct. 1945
519	Handley Page Halifax III	Aug. 1945–May 1946
520	Handley Page Halifax III	Apr. 1945–Apr. 1946
521	Handley Page Halifax VI	Dec. 1945–Apr. 1946

Squadron	Aircraft	Date
527	Vickers Wellington X	Apr. 1945–Apr. 1946
544	Vickers Wellington IV	Oct. 1942–Mar. 1943
547	Vickers Wellington VIII	Nov. 1942–May 1943
	Vickers Wellington XI	Apr. 1943–Nov. 1943
	Vickers Wellington XIII	Oct. 1943–Nov. 1943
570	Armstrong Whitworth Albemarle I	Nov. 1943–Aug. 1944
578	Handley Page Halifax B. Mk III	Jan. 1944–Mar. 1945
612	Vickers Wellington VIII	Nov. 1942–Jun. 1943
	Vickers Wellington XII	Apr. 1943–Jun. 1943
	Vickers Wellington XIV	Jun. 1943–Jul. 1945
614	Handley Page Halifax B. Mk III	Mar. 1944–Mar. 1945
620	Short Stirling III	Aug. 1943–Feb. 1944
622	Short Stirling III	Aug. 1943–Dec. 1943
633	Short Stirling III	Aug. 1943–Dec. 1943
624	Short Stirling IV	Jun. 1944–Sept. 1944
640	Handley Page Halifax B. Mk III	Jan. 1944–Mar. 1945
	Handley Page Halifax B. Mk VI	Mar. 1945–May 1945
644	Handley Page Halifax A. Mk III	Aug. 1944–Nov. 1945
	Handley Page Halifax A. Mk VII	Mar. 1945–Sept. 1946
684	Bristol Beaufighter VI, X	Aug. 1945–Oct. 1945

8

THE HERCULES AND THE BOMBER COMMAND MUSEUM OF CANADA

There are two very special aviation museums in Canada that are connected with the Bristol Hercules engine. One is the Bomber Command Museum of Canada (BCMC) in Nanton, Alberta. Nanton is a farming community nestled between the prairies and the foothills of the Rocky Mountains. The other is the National Air Force Museum (NAFM) located on the west side of Canadian Forces Base Trenton, Ontario. The base is on the north side of Lake Ontario east of Toronto. What is the connection? The NAFM rescued and restored a Handley Page HP 63 Halifax A Mk VII from Lake Mjosa, Norway, which is now displayed at the museum. The BCMC is presently rescuing a Handley Page Halifax from the sea off Sweden and plans to restore it for display at the museum. In advance of that display, the BCMC is unique in possessing a ground running Bristol Hercules 734 engine on a stand, which is started at all major public events. Both museums are a credit to their organisations and the army of volunteers who have brought the museums to such a high standard.

I am very familiar with the BCMC having researched all my previous books (see www.amberley-books.com) with Dave Birrell, Archivist and Librarian, Greg Morrison, the chief Avro Lancaster engineer, and Doug Bowman, museum photographer. For this

The Hercules and The Bomber Command Museum of Canada

book I had the pleasure of working with Karl Kjarsgaard, the project leader on the Halifax recovery and restoration. Kjarsgaard is also responsible for the ground running Hercules engine. I would like to digress for a moment to recount the fascinating history of the BCMC before addressing the Bristol Hercules component within the present-day activities of the museum. A small prairie town has built a memorial to those who lived and died for our freedom as members of the Royal Canadian Air Force and Royal Air Force Bomber Command in the Second World War. Initially, the focus of the museum was on preserving, then restoring, a ground running Rolls-Royce engined Avro Lancaster to attract visitors to the small community. Other aircraft, thirteen at last count, and exhibits were collected along the way, culminating today in the ongoing rescue and restoration of a Bristol Hercules engined Handley Page Halifax.

The BCMC sits on the edge of the town of Nanton alongside the highway. The apron is so small that traffic on the highway has to be diverted to one lane to allow the aircraft to be towed into position for the engine run-up displays. A temporary wire barrier is put in place to keep spectators back from the aircraft. Needless to say, not many people drive by without stopping to look at the Lancaster, especially when it has its engines running. There are ambitious plans underway to enlarge the active display area by relocating it beside the hangar.

Nanton is seventy-five kilometres (forty-five miles) from the major city of Calgary, in a predominately agricultural area. During the Second World War it was surrounded by active stations of the British Commonwealth Air Training Plan (BCATP) due to its terrain and good flying weather. Some twenty-eight kilometres (seventeen miles) north of Nanton, Elementary Flying Training School No 5 at High River operated for virtually the entire duration of the war. At the Vulcan Station east of Nanton, the BCATP operated Service Flying Training School No 19, and during the early years of the war, Flying Instructor School No 2. Another Service Flying Training School, No 15, was located at Claresholm, forty kilometres (twenty-four miles) south of Nanton.

The Hercules: The Other Engine That Helped Win the War

Nanton did not have any real reason to establish an aviation museum. It was a way for some citizens to draw attention to their community and at the same time preserve some of the surrounding aviation heritage. Three were the prime movers: George White, Howie Armstrong, and Panton Garratt. White was a local farmer and rancher who lived on his family land, which they had homesteaded in the late nineteenth century. There were flying training stations in the vicinity, and he was used to seeing training aircraft in the skies above his family's ranch during the war years. He did in fact obtain his flying licence but never had the chance to take it further; there was not much time left in the day after farmwork was done. White had the initial idea to take a war surplus aircraft and put it on display as a memorial and as a tourist attraction to get the highway traffic to stop and perhaps put some badly needed money into the local economy.

He had instant support. Armstrong was a Nanton entrepreneur who owned Armstrong's Department and Variety store. He championed the small town in any way he could, and one was to label the piped-in town water as 'Canada's Finest Drinking Water' and provided a tap for passing travellers. Nanton Water Ltd later became one of the first companies in Canada to bottle and sell drinking water. Garratt operated the McKeague and Garratt Hardware store, an invaluable source of products which would be used to display and, more importantly, preserve the aircraft exposed to the harsh prairie weather.

In 1960, several surplus North American B-25 Mitchell bombers were being flown to the relatively close Claresholm, Alberta airfield for disposal. White wrote to Crown Assets and enquired as to the cost of one of these aircraft. The selling price was $2,500, too much for the group, so they enquired if there was anything else available. They were told that some Lancasters would be put up for tender at a nearby airfield, a long-term storage facility near Vulcan, Alberta. An offer of $513 (£286) was accepted and the group now owned an Avro Lancaster (FM159), albeit in a rural airfield, which, as it turned out, was a blessing in disguise.

The Hercules and The Bomber Command Museum of Canada

On 12 February 1959, FM159 had been flown by a civilian crew from Calgary to Vulcan. It was an ex RCAF 407 Squadron aircraft. By March the engines and propellers had been taken off and along with the fuselage had been put in storage. The Nanton group acquired the aircraft on 11 August 1959. Now the problems started. The main landing gear was too wide for the rural gravel roads and then there was the matter of the numerous telephone poles. The eventual answer was to tow the aircraft by the shortest route across the fields for twenty-eight kilometres (seventeen miles).

Permission was obtained from all the farmers involved with the proviso that the aircraft move would have to wait until after the crops were off the fields. Planning, volunteers, and equipment were all readied for the big day. Archie Clark, who had experience with towing large equipment, was chosen to do the towing. The tail wheel was secured on the truck flatbed and two logging chains were attached to the main landing gear for security. The aircraft was then towed backwards across the prairie fields.

Ditches were filled and excavated, wire fences removed, and rebuilt as the armada of volunteers moved along with the Lancaster. The route included towing through a ford in the Little Bow River and crossing a railway line before getting to the highway for the final route into Nanton. The telephone lines required that someone rode on top of the aircraft and helped ease the fuselage underneath the wires. The strange procession came to a halt for the night at the Canadian Pacific Railway's tracks just three kilometres (two miles) from Nanton. The permission to cross the tracks was not until the next morning. A guard was set up to protect the aircraft. The next morning the tow was completed down the highway for the triumphant entry into town on the 28 September 1960. The Lancaster was parked by Highway No 2 and immediately proved itself to be an eye opener for the passing traffic.

The fuselage was totally complete as the day it had been put in storage. Unfortunately, sitting on the ground at the edge of town exposed it to passing souvenir hunters and vandals. By the

Autumn of 1961, in spite of the best efforts of the volunteers, FM159 was a gutted shell of its former self. Broken Perspex, stolen cockpit instruments, damaged turrets and torn fabric control surfaces turned the once proud Lancaster into a sorry sight of dereliction; had the effort of attracting attention to the town been all for nought?

1962 saw a change of fortune for the aircraft. The original three preservers mounted a campaign to save the aircraft from total destruction. Engines and propellers had been purchased earlier and in December the work began to mount them to the fuselage to form a complete-looking aircraft. In 1963, the aircraft was securely mounted on steel mounts using its main landing gear. The tail wheel was raised to put the aircraft in a horizontal position to prevent easy access to the insides of the aircraft. Aluminium windows were fabricated later to prevent further damage from bird droppings and the aircraft was painted in wartime colours.

Nose art was added, naming the aircraft B for Bull Moose. Bomb tally was added showing fifty missions flown and three enemy fighters shot down. A sign was constructed to indicate the part that the Lancaster bomber had played in the defeat of the Nazis. Volunteers, under the guidance of Ray McMahon, continued to maintain FM159 as it sat beside the highway for the next twenty years.

In 1985, an article in the *Nanton News* by Herb Johnson drew attention to the plight of Lancaster FM159. Vandalism and the prairie climate were continuing to take their toll on what had now become a very desirable historic aircraft to interested persons and organisations outside the Nanton community. The town council decided to ask White, an instigator of the Lancaster purchase, to see if there was enough local interest to form a society to look after the old aircraft. They were overwhelmed by the response. Assessment began to see where to get the expertise to look after the aircraft and how to fund further preservation by housing the aircraft in a permanent building. The Executive Summary of the proposal stated: 'The most important aircraft in the museum will be the well-known Nanton Lancaster Bomber which will be restored to its wartime configuration and to a 'taxiable' status.'

The Nanton Lancaster Society (NLS) was subsequently formed. In spring 1986 the members began an inspection of the aircraft. The first hurdle was to get the padlock open on the crew door! By design the steel mounting of FM159 on its landing gear preserved the fuselage intact as opposed to the cutting and welding of a pedestal-mounted aircraft. Although stripped of its interior equipment and instruments, the aircraft had remained clean thanks to Harry Dwelle sealing the broken Plexiglas with metal sheeting.

In 1987, the general public had access to the aircraft, which has continued to the present day. Over 700 people toured the aircraft during the first weekend. The connection of the Avro Lancaster to the visitors surprised the members of the society. Everything from 'Uncle Harry flew these in the War,' to 'My Mum worked on these in Malton, Ontario,' to 'I have a few hours in the Lancaster.'

In 1987, Jon Spinks, an aviation aficionado who was an expert on finding parts for the Lancaster fully restored the vandalised pilot's instrument panel. Spinks also encouraged the Society to expand their horizons and start collecting other aircraft of the BCATP before they were all gone for scrap. Perhaps the most important decision of the NLS at this time was the decision to honour the people associated with Bomber Command. The people who built the aircraft, the people who maintained the aircraft, the people who trained the people to fly the aircraft, and the people who flew the aircraft would all be honoured, quite a commitment from a farming community of 2,000 people on the eastern slopes of the Canadian Rocky Mountains.

Millions of visitors over the years from far and wide knew about the 'Nanton Bomber' and it featured in many a family photograph. It had achieved what White, Armstrong, and Garratt hoped it would, publicity for their small prairie town. Nanton became the little town that put itself on the tourist map with one old aircraft towed across some fields. FM159 would have

The Hercules: The Other Engine That Helped Win the War

further exciting transitions in its future, but what of its past? Questions, such as did it cross the Atlantic and take part in sorties over enemy territory, what happened to it after hostilities ceased, would be answered by careful research and the stories of veterans who had flown in FM159 and later shared their experiences with the museum. Exact records are not available but FM159 Construction Number 3360 was built at Victory Aircraft, Malton, Ontario, probably in the latter half of May 1945. It was number 360 of the 430 Lancasters built in Canada. The momentum of the war effort (Canada and Japan were still at war) caused FM159 to be flown to England after VE (Victory in Europe) Day between 29 and 31 May 1945. The aircraft was saved from being scrapped due to its low mileage and the possibility of it joining the 'Tiger Force' for operations in the continuing war in the Far East. Two months at No 20 Ashton Down and No 32 St. Athan Maintenance Units and the aircraft was flight tested and flown back to Canada on 30 August 1945 to RCAF Station Scoudouc, New Brunswick.

Subsequently, FM159 was flown to Yarmouth, Nova Scotia, for storage as after VJ Day on 14/15 August. There was no longer a Tiger Force requirement. Fear of salt air corrosion prompted the aircraft to be moved in the autumn to former BCATP airfields in the drier climate of Alberta. FM159 was flown to the former No 2 Flying Instructor School at RCAF Station Pearce, which is Northeast of Fort Macleod. The aircraft were either properly stored in the hangars, or could be purchased for parts and pieces that could be used with ingenuity by the local farmers.

FM159 once again was saved from this indignity and flown to Fort Macleod in March 1946. It remained there for five years until 1951. In August of that year, it was flown back east to No 6 Repair Depot at RCAF Station Trenton, Ontario. Finally in March 1953 FM159 came back to life again being flown to Malton, Ontario for overhaul and conversion by De Havilland Aircraft Company of Canada to a Maritime Reconnaissance (MR) version. About 100 of the returned Lancasters continued to serve the RCAF, nearly 200 were scrapped. The conversion included the installation of a co-pilot position, radar installation, sonobuoy capability, additional fuel, and observation blister windows. Seven months later in October 1953 FM159 began flying duties with No 103 Search and Rescue Unit based at Greenwood, Nova Scotia.

By January 1955 it was scheduled for radar system upgrades by Fairey Aircraft at Dartmouth, Nova Scotia. In June the aircraft had been assigned to No 407 (MR) Squadron at RCAF Station Comox, British Columbia. The aircraft had left the Atlantic Ocean and now would serve the RCAF on the Pacific Ocean. Shortly thereafter, FM159 left on an Arctic ice reconnaissance patrol, an annual tasking called 'Nanook'. It was part of the ongoing survey of sea routes to document when the routes were available for Distant Early Warning Line resupply ships.

The winter resupply of the North American Air Defence Command radar sites in the Arctic by aircraft was expensive, and the sooner the ships could get through the ice the better. During this deployment FM159 nearly became a victim of the notorious Arctic weather. While on patrol out of Resolute Bay, Cornwallis Island, Nunavut, low ceilings and fog moved in over the whole area. Despite proceeding to Thule in Greenland and examining every possibility of a landing airfield, the only open landing area was at Alert, Ellesmere Island, Nunavut, the northernmost point in Canada. It is just 817 kilometres (508 miles) from the North Pole. FM159 eventually landed there, nearly out of gas and nowhere to go, after a flight of twelve hours and twenty-five minutes. The number four engine failed on the runway after landing, the fuel tank was empty. That is how close FM159 came to being an aircraft accident statistic.

The aircraft continued to serve with No 407 (MR) Squadron and was even tasked with NORAD penetration exercises in April 1956. Intercepted by RCAF Avro CF-100s it continued the mission for training purposes and did a simulated bomb run on Spokane, Washington, US. It succeeded in its mission prior to interception by USAF F-86 Sabres and so earned the right to display a small 'bomb' on the side of the fuselage, mission

The Hercules: The Other Engine That Helped Win the War

accomplished. Squadron service continued until December 1958, when it was flown to Calgary and once again parked, destined for an unknown future.

Two months later FM159, along with other No 407 Squadron Lancasters, was flown to RCAF Station Vulcan, Alberta. The Lancaster aircraft's service to No 407 Squadron was completed in May 1959 to be replaced by the Lockheed P2V-7 Neptune. This is where White, Armstrong, and Garratt became involved with FM159. So what happened after the 'Open Bomber Days' in 1987 to get FM159 to the ground running example it is today? The public showed that they were interested in the project by their attendance and the NLS was gathering strength. A blizzard in 1989 tore the port elevator off and left it dangling from the horizontal stabiliser. This really drew attention to the fact that the forty-five-year-old aircraft would not last much longer if left exposed to the prairie climate. A museum building was required to house the aircraft and provide space for educational displays. The town of Nanton provided an interest-free loan, which, together with numerous individual donors, resulted in the aircraft being removed from its supports and towed by tractor in 1991 into the newly built hangar. Calling it a hangar stretched the imagination; it was a cover for the Lancaster that was sitting on a donated gravel floor.

Prior to the Lancaster getting a roof over its head it had been decided by the NLS in 1990 to dedicate the Lancaster to a local unknown hero who had served with the RAF Bomber Command. This was done to publicise and recognise all who had served in Bomber Command during the Second World War. This provided a focus for the museum volunteers; they were not restoring any Lancaster, it now had a name and a personal story behind it.

The Nanton Lancaster was dedicated in a ceremony on Friday 27 July 1990 to the memory of S/L Ian Willoughby Bazalgette VC DFC RAF VR, a local Alberta boy. Those in attendance included his sister, Mrs Ethel Broderick (née Bazalgette) and two of his former crew, Charles 'Chuck' Godfrey DFC, Bazalgette's Wireless Operator and George Turner, Bazalgette's Flight Engineer. Hamish

The Hercules and The Bomber Command Museum of Canada

Mahaddie DSO DFC AFC and Bar had selected Bazalgette, and others, for the important role of being a Path Finder. During his banquet speech it was obvious that he felt very deeply for those crew who had paid the ultimate sacrifice doing the very important job of marking the target. A dedication plaque was unveiled by Mrs Broderick to officially name the Nanton FM159 as the 'Ian Bazalgette Memorial Lancaster'.

Bazalgette completed Operational training on the Vickers Wellington and was posted to No 115 Squadron RAF where he won the Distinguished Flying Cross on 1 July 1943. The citation read: 'This officer has at all times displayed the greatest keenness for operational flying. He has taken part in many sorties and attacked heavily defended targets such as Duisberg, Berlin, Essen and Turin. His gallantry and devotion to duty have at all times been exceptional and his record commands the respect of all in his squadron.'

Bazalgette has a further connection to the present-day museum. He completed his operational tour flying the Bristol Hercules VI engined Avro Lancaster BII.

The following is the citation accompanying the awarding of the Victoria Cross to S/L Ian Bazalgette as published by the London Gazette on the 17 August 1945:

On August 4, 1944, Squadron Leader Bazalgette was master bomber of a Pathfinder squadron detailed to mark an important target at Trossy St. Maximin for the main bomber force. When nearing the target his Lancaster came under heavy anti-aircraft fire. Both starboard engines were put out of action and serious fires broke out in the fuselage and starboard mainplane. The bomb aimer was badly wounded. As the deputy master bomber had already been shot down the success of the attack depended on Squadron Leader Bazalgette, and this he knew. Despite the appalling conditions in his burning aircraft he pressed on gallantly to the target, marking and bombing it accurately. That the attack was successful was due to his magnificent effort. After the bombs had been dropped the Lancaster dived practically out of control. By expert airmanship and great exertion Squadron Leader Bazalgette regained control, but the port inner engine then failed and the whole of the starboard mainplane became a mass of flames. Squadron Leader Bazalgette fought bravely to bring his aircraft and crew to safety. The mid upper gunner was overcome by fumes. Squadron Leader Bazalgette ordered those of his crew who were able to leave by parachute to do so. He remained at the controls and attempted the almost hopeless task of landing the crippled and blazing aircraft in a last effort to save the wounded bomb aimer and helpless air gunner. With superb skill and taking great care to avoid a small French village nearby, he brought the aircraft down safely. Unfortunately, it then exploded and this gallant officer and his two comrades perished. His heroic sacrifice marked the climax of a long career of operations against the enemy. He always chose the more dangerous and exacting roles. His courage and devotion to duty were beyond praise.

Dave Birrell, the BCMC archivist and librarian, pointed out that there were two errors in the citation, that Baz was not the Master Bomber on that raid and that fires did not break out in the fuselage. Perhaps a case of the finer details being lost in the confusion of war. The following day, 5 August 1944, S/L I.W.Bazalgette was permanently awarded his Path Finder Force badge. Baz's parents donated his medals to the Royal Air Force Museum at Hendon, London, where they are on display.

Birrell was one of the original founding Directors of the NLS. A former geophysicist in the oil and gas industry, he changed career to become a schoolteacher and assumed a job at the local Nanton school. He joined the NLS in 1986 as part of his becoming involved in the local community, anticipating 'cutting grass underneath the Lancaster in summertime'. He was initially responsible for treasurer duties, collecting memberships and accepting donations. The town Tourist Information building displayed Lancaster memorabilia. Birrell has been involved with the museum for nearly thirty years: 'My time with the museum has been multi-faceted, which has held my attention and prevented me from losing interest.' Birrell was awarded the Queen's Jubilee Medal in 2011 for his service to the museum.

In 2000, a suitable room became available in the 'lean to' portion of the hangar which included the front entrance. Shelves were erected, filing cabinets filled and computers installed and Birrell had created a library and archives section for reference and research.

By 1991, with the Lancaster now dedicated to S/L Bazalgette, the focus of the NLS slowly shifted to Bomber Command itself. Related aircraft and displays were under development. The basic hangar was expanded to house a library and archives, parts storage area, and office. Emphasis was placed on artifacts, aviation art, and interpretive information displays. In 2003, big doors were added to the hangar, which enabled the Lancaster to be pulled out on to the apron.

The starting of #3 engine in 2005 coincided with the museum's Memorial Wall project built during the summer of 2005. It was dedicated at a ceremony on 20 August 2005. The twelve metre (forty-one foot) Memorial is made of five polished, black granite panels. Four of these panels each have about 1,600 names engraved per side. The central panel includes the name and purpose of the Memorial an image of a bomber crew, and the Bomber Command crest. It is located on the front lawn of the Bomber Command Museum. The Memorial's 10,872 engraved names include Canadians from every part of the country that served in Bomber Command and were killed in the line of duty. In 2010, the name of the museum was changed to the Bomber Command Museum of Canada to better reflect its scope and objectives. Two major expansion projects in 1997 and 2007 resulted in space for thirteen aircraft in a 3,437 m^2 (37,000 ft^2) hangar. The museum has evolved into a facility that restores aircraft to runnable, taxi status. The future plans include a 3,121 m^2 (33,600 ft^2) hangar with a run up display area of 5,268 m^2 (56,700 ft^2).

The restoration of the Lancaster to ground running condition took many hours of volunteer labour over many years, Finally,

The Hercules: The Other Engine That Helped Win the War

on 24 August 2013, the sound of four Merlin engines were heard throughout Nanton. BCMC now had a world class exhibit for enthusiasts to come to from all over the world. The Lancaster performs its four-engine ground run during scheduled events during the year.

It is into this vibrant, successful museum scenario that the H57RC, Halifax 57 Rescue (Canada) project appeared under the leadership of Karl Kjarsgaard. The 57 is the Handley Page company designation for the first model of Halifax. The Halifax will be a perfect companion for the Lancaster. Two ground running four-engine, heavy bombers together, now that will be some accomplishment – not to mention the combined sound of four Rolls-Royce Merlins and four Bristol Hercules engines.

Kjarsgaard started his flying career with Wardair, a Canadian airline, flying the De Havilland DHC-6 Twin Otter based in Yellowknife, Northwest Territories. Then in 1972 he upgraded to the Bristol 170 freighter which had the Bristol Hercules engines. Who could have foreseen that fifty years later he would be restoring Hercules engines to ground running condition for the BCMC?

Towards the end of a successful career with Canadian Pacific Airlines, Canadian Airlines, and Air Canada Kjarsgaard decided to indulge his passion for Second World War aircraft. As a student pilot, he had been taught by old Second World War pilots who told him about their aircraft and combat operations. He learned that many of the Canadian squadrons flew the Handley Page Halifax almost exclusively during the war. It became apparent to him, listening to all the stories, that the early Halifax lacked performance with the Rolls-Royce Merlin engines, but the Halifax was transformed with the Hercules.

In Toronto he helped start a group called the Halifax Aircraft Association to preserve the history of the aeroplane. Kjarsgaard wanted to take it one step further and he searched world-wide and found a Halifax at the bottom of a lake in Norway which was recovered in 1995. When they got Halifax NA337 from Norway to Trenton, Ontario, he wanted to create an organisation to procure all the parts needed for the rebuild. A friend, Ian Foster, said that we would start the parts research group and call it the Halifax 57 Rescue. The name came from the fact that every ten-digit part number for the Halifax aircraft started with 57. When the Trenton rebuild was over, Kjarsgaard took over and renamed it H57RC (Halifax 57 Rescue (Canada)) as a separate legal charity in 2004.

In the years 1995 to 2005 Kjarsgaard had dealt with the BCMC extensively during the Trenton rebuild and admired the museum volunteers' enthusiasm. On retirement he decided to move and work with the BCMC and convinced the innovative BCMC Board that a partnership between the BCMC and H57RC would result in a ground-running Halifax to display alongside the Lancaster.

It would take many years to find and rebuild a whole Halifax aircraft, but a ground-running Bristol Hercules was closer within reach. By 2011 he had succeeded.

The restored engine on the running display stand is the Bristol 734 model. The project now has eight engine blocks in various states of disrepair; the best can be restored to running condition and the rest will be salvaged for spare parts. Two engines are basically spare parts engines found in Malta. Kjarsgaard was aware that he needed British tools to work on the Hercules and knew that there was an airline, Hawk Air, in Terrace, British Columbia, that had at one time flown the Bristol Freighter. When he enquired about the tools, he was told that not only could he have the tools but there were four Hercules engines available; just take them away as a donation to the museum. The deal, and it was a deal, included the requirement to take all the spare parts also, new and used. The best of the four engines was in running order.

The final two engines of the collection, almost ready to run, fully inspected, and rebuilt, were from Leeds in the United Kingdom. They are from a Handley Page Hastings aircraft. The engine models in inventory are four 734s, two 215s, and two 230s. The naming convention is military engines up to 230

The Hercules: The Other Engine That Helped Win the War

and the civilian engines are the military number plus 500. For example, the military version is 234 and the civilian version is 734 as used on the Bristol 170 Freighter. Thinking back to his time on the Bristol Freighter he remembered being told, 'They are powerful, don't flood the engine, and they will run forever if you are careful with them.'

When the engines arrived at Nanton, Kjarsgaard 'learned by osmosis' from the guys who got it running. The engines do differ from the wartime Hercules engines in the cylinder head shape and the throttle body fuel injection system.

The gifted lead machinist was Dan Hawken of High River who got the first Hercules ready to run and worked alongside Derek Squire, a car mechanic, with Kjarsgaard as the 'gopher'. However, all were learning from Dan as the work progressed. A primary job of the gopher was to find the manuals and the information available on the various complicated components of the engine. Dan went through the engine painstakingly and made sure each component was functioning properly.

Kjarsgaard described how he would 'take a wooden propeller from an Anson to England and trade it to a Brit, and he would give me some piston rings for a Bristol Hercules'. He would make more than twenty trips to Britain to accumulate the necessary parts and use a hire van to pick up a propeller hub here, sets of four piston rings there, and spark plugs somewhere else. Hundreds of Hercules and Halifax parts were acquired.

A trailer company in Fort Macleod offered to build a road certified engine-run trailer and donate it to the museum. The oil and gas tanks were moved over to the new trailer and now it was easy with a trailer hitch to move the 907 kg (2,000 lb) engine outside for events.

Kjarsgaard assumed more and more of the engineering work as the years progressed. He worked with Derek Squire to replace the blown supercharger on No 1 engine. Fortunately, he had Paul Hawkins in BC and Patrick Smart in the UK for guidance. The replacement supercharger was scavenged from one of the Malta Hercules engines.

The Hercules and The Bomber Command Museum of Canada

To move the trailer into position there is a ball hitch on the museum's forklift. When in position two jack screw posts are extended to the ground to give support and chocks are positioned in front of and behind the wheels on both axles. The entire bottom of the yellow trailer is a drip tray underneath the engine. It has a drain plug to remove accumulated oil. There are two members of the run crew. One operator is on the rear of the engine controlling the throttle and mixture and is partially protected from the propwash by the engine. The other operator, the crew chief, is at the hand-carried control panel. This panel is at the end of a 6 m (20 ft) cable and controls the Master Switch. This crew chief gives hand commands to the operator controlling the engine. Ear protectors are worn as the engine produces 100 decibels at low to medium power. This, of course, is part of the crowd appeal, in addition to the smoke and backfiring during the start.

The instruments on the engine firewall visible to the operator are:

1) Electric fuel pump pressure.
2) Engine driven fuel pump pressure.
3) Tachometer.
4) Oil pressure gauge.

The switches on the control panel are:

1) Master switch.
2) Electric fuel pump.
3) Primer valve switch to allow fuel to enter supercharger chamber.
4) Starter switch.
5) Magneto switches, Left and Right.

In addition, there is a hand-held infra-red gun to read cylinder head temperature.

When starting, it's one minute on, 20 seconds off, maximum of three attempts. A starting problem, if it occurs, is mostly associated

The Hercules: The Other Engine That Helped Win the War

with too little or too much fuel in the cylinders. Idling, the throttle venturis are set at twenty-two degrees, and not moved for a minute or so. The venturi scale is next to the throttles. The idle tends to be rough as the cylinders clear out. The RPM is around 900 to 950. The throttles are then increased to around 1,200 RPM, as Kjarsgaard said, 'Seems to clear its throat'. Once 100°C cylinder head temperature is indicated on the infra red gun, the RPM is usually increased to 1,500–1,800 RPM. The absolute maximum is 2,000 RPM. Keeping in mind that the takeoff RPM was 2,900! There are normally two runs per BCMC event.

One of the rebuilt engines brought from Leeds will be the next Hercules installed on the trailer for engine running duties. The running Hercules is the perfect advertisement to draw attention to the fact that the museum is embarked upon a project to restore a Handley Page Halifax to ground running condition.

This Handley Page Halifax HR871 was flying with No 405 Royal Canadian Air Force Pathfinder Squadron on the 3 August 1943 when it was hit by lightning and two of its engines failed. It diverted to neutral Sweden where the seven crew all managed to bail out successfully. No 405 Squadron itself was a truly international squadron with members from Britain, Australia, and America flying for the RCAF squadron. The aircraft crashed in the Baltic Sea fifteen kilometres (nine miles) off the coast of Sweden and sank in 14 metres (40 feet) of brackish water. It was discovered by a Swedish diving group, the Swedish Coast and Sea Center (SCSC) who found it mostly covered in sand with some engine and fuselage pieces visible on the sea floor.

The Swedish Government gave permission to Halifax 57 Rescue for the Canadian aircraft to be brought back to Canada. The *River Thames*, a refurbished British tugboat, is a fully equipped operational diving and sand clearing headquarters for the project. It has a workshop, pump for high pressure sand clearing, and a diving air compressor. In addition the main portion of a Halifax wing and other airframe parts are being repaired and rebuilt from factory blueprints by Scott Knox of Knox Tech Inc., Arnprior, Ottawa, Ontario. The hunt for parts

The Hercules and The Bomber Command Museum of Canada

continues. A Halifax Perspex nose was recently discovered in Cullen, Scotland.

The final aim of all this expensive recovery work is to restore a Halifax with four Hercules engines. The completed Handley Page Halifax with its Bristol Hercules engines will be a fitting tribute to the RCAF Canadian crews of RAF Bomber Command who sacrificed their lives so we may enjoy our freedom eighty years later.

To contribute to this ongoing project please see online: FUNDRAZR 417498.

CONCLUSION

It has been an amazing literary exploration going back nearly one hundred and twenty years to when a young Roy Fedden completed his apprenticeship and immediately started designing a car. Though a social disappointment to his upper crust family as he embarked on a career as an engineer, even at this early stage of his career Fedden was a very focused individual and it did not seem to bother him. Fedden also understood that he could not achieve great success without a a good team around him.

He was fortuitous in his choice of a working partner in a company that was ready to take on the aviation challenges of the time. Fedden worked with 'Bunny' Butler and they combined to produce the Straker Squire car. Their entry into the world of aero engines was the challenge to see if they could fix the problems with the Curtiss OX-5 engine. They did so and thus established a good reputation with the British government.

Fedden decided on the air-cooled, radial engine as the way to go and just after designing the Jupiter engine he found himself working for a new company. This company, Cosmos Engineering, also foundered and Fedden was left holding the Fishpond Works for the Receiver. This should have been the end of Fedden's ambition in the aviation engine world, but luck was on his side.

Conclusion

The Bristol Aeroplane Company were coerced by the government into taking over Cosmos engineering. Fedden quickly established an engineer apprenticeship program and the big break came when he re-engineered the Jupiter and exhibited at the Paris Airshow. His reputation, and more importantly the Bristol Aeroplane Company's, was intact in the aftermath of the First World War and he could now concentrate on developing the air-cooled radial engine.

Partly through government intervention post-war, Bristol now had the full package, aircraft and engine, thanks to Fedden and Cosmos Engineering. Bristol would produce thirty aircraft, racers, fighters, bombers, and transports, using its own engines and others to produce biplanes and the now emerging monoplane design. The Jupiter engine was helping to build the company by being built under license. By 1936, Bristol was a confident, viable company ready for Fedden's newest engine, the Bristol Hercules.

Fedden was emboldened to break convention and develop the new concept of the sleeve-valve engine rather than the standard poppet-valve engine. The sleeve-valve engine did away with the complicated valve mechanism of the poppet-valve engine. The push rods, rocker arms, tappets, valve springs, and valves were all been replaced by a sleeve with openings in it.

Over a seven-year development period Fedden created a series of nine, fourteen and eighteen-cylinder engines. The successful sleeve-valve engines, the Perseus engine in 1932, followed by the Aquila engine in 1934, would lead to the development of the famous Bristol Hercules engine. Fedden took two banks of the cylinders from the Perseus engine and staggered them together to make the eighteen-cylinder Hercules engine. The heart of the engine is the extraordinary sleeve-driving mechanism on the front section of the crankcase with fourteen cranks linked to the individual sleeve, long spindles passing through to the rear. Gears upon gears have layshafts attached.

It was a 1936 marvel. The 700 series of Hercules engines would produce 2,040 hp at takeoff. The initial horsepower rating had increased by a third by the end of hostilities. Fifteen manufacturers would use the Hercules in an engine testbed, prototype, or production aircraft. These companies would include the giants of the British aviation industry. Avro, Handley Page, Short, and Vickers would design their wartime aircraft around the performance of the Hercules engine. Other British companies, who would do the same were Blackburn, Bristol, Fairey, Folland, and Saunders-Roe. These companies had the unenviable task of deciding what engine would work best for their project, the air-cooled radial Hercules or the liquid-cooled inline Merlin engine.

The Hercules was not limited to British companies and saw varying levels of service with American, Northrop; French, Nord, Breguet; Spanish, Casa; and Holland, Fokker; The Hercules would continue post-war for a number of years in the transport role, The Handley Page Hastings, being an example, which served with the RAF until 1977. Similarly, the Nord Noratlas electronic warfare version would serve the Armee de l'Air until 1989. Descendants of these English and French aviation companies, British Aircraft Corporation and Sud Aviation, would work together to create Concorde.

In the Second World War the main Hercules powered fighter was the Bristol Beaufighter in all its forms, including service with the RAF Coastal Command. The core of the RAF bomber force comprised the Avro Lancaster BII, variants of the Handley Page Halifax, the Short Stirling, and the Vickers Wellington, all with the Hercules engine. The flying boats of Coastal Command, the Short S26 and Short Seaford, performed sea duties against the enemy while the Short Solent was a passenger aircraft. The transport aircraft included the RAF Handley Page Hastings, Vickers Valetta, and Vickers Varsity. The Bristol Freighter, Handley Page Hermes, and Vickers Viking were part of the post-war civilian airline market.

On a personal note, I was amazed to see how many Canadian airlines had used the Bristol Freighter in remote areas, over mountainous terrain, in the Canadian Arctic, and on unprepared

Conclusion

surfaces. One of them, Hawkair of Terrace, British Columbia, generously donated Hercules engines, special tools, and spare parts to Kjarsgaard and the Bomber Command Museum of Canada. Reputedly the last flight by a Bristol Freighter was a Mk 31 donated by Hawkair to the Reynolds-Alberta Museum in Wetaskiwin, Alberta in 2004.

BIBLIOGRAPHY

Air Ministry, Ministry of Information, *Bomber Command* London: Harrison & Sons Ltd. 1941
Allward, Maurice, *Hurricane Special*, Shepperton: Ian Allan Ltd 1975
Armitage, Michael, *The Royal Air Force; An Illustrated History*, London: Brockhampton Press 1996
—— *The History of the Royal Air Force*, London: Weidenfeld & Nicholson Ltd. 2000
Ashton, J. Norman, *Only Birds and Fools*, Shrewsbury: Airlife Publishing Ltd. 2000
Barnes, C. H., *Bristol Aircraft since 1910*, London: Putnam & Company Ltd 1964
Barris, T., *Behind the Glory*, Toronto: Macmillan Canada 1993
Bercusson, T. J. *Maple Leaf against the Axis: Canada's Second World War*, Toronto: Stoddart 1995
Bingham, Victor F., *Handley Page Hastings & Hermes*, GMS Enterprises 1998
Birrell, D. *Baz: The Biography of Ian Bazalgette VC*, Nanton: The Nanton Lancaster Society 2014
—— *People and Planes: Stories from the Bomber Command Museum of Canada*, Nanton: The Nanton Lancaster Society 2011

Bibliography

—— *FM159: The Lucky Lancaster,* Nanton: The Nanton Lancaster Society 2015

—— *Johnny: Canada's Greatest Bomber Pilot,* Nanton: The Nanton Lancaster Society 2018

—— *People and Planes: Stories from the Bomber Command Museum of Canada,* Nanton: The Nanton Lancaster Society 2011

Bishop, P. *Bomber Boys: Fighting Back 1940–1945,* UK: Harper Collins Publishers 2008

Bowman, Martin W., *100 Group (Bomber Support): RAF Bomber Command in World War II,* Barnsley:Pen & Sword Books Limited 2006

Bowman, Martin W., *100 Group (Bomber Support): RAF Bomber Command in World War II,* Barnsley: Pen & Sword Books Limited 2006

'Bristol' Aircooled Radial Engines, Bristol Aeroplane Engines Limited: manual, Montreal

Bowyer, Chaz, *History of the RAF,* London: Hamlyn Publishing Group Ltd. 1985

—— *RAF Operations 1918–1938,* London: William Kimber & Co. Limited 1988

Braddon, Russell, *Cheshire V. C.,* London: Evans Brothers Limited 1954

Chisholm, Anne & Davie, Michael, *Beaverbrook A Life,* London: Pimlico 1993

Chorlton, Martyn, *The RAF Pathfinders: Bomber Command's Elite Squadrons,* Newbury: Countryside Books 2012

Chorlton, Martyn, *Company Profile 1910–1959: Bristol,* Cudham 2014

Christie, C. A., *Ocean Bridge: History of RAF Ferry Command,* Toronto: University of Toronto Press 1997

Delve, K. *RAF Bomber Command 1936–1945: An Operational and Historical Record,* UK: Pen & Sword 2006

Duke, Neville, *Test Pilot,* Plymouth: Latimer, Trend & Co. Ltd 1953

Edwards, Richard and Edwards, Peter, *Heroes and Landmarks of British Military Aviation: From Airships to the Jet Age,* Barnsley: Pen & Sword Books Limited 2012

Embry, Sir Basil, *Mission Completed,* London: Methuen & Co Ltd 1957

Falconer, J., *Bomber Command Handbook:1939–1945,* Stroud: Sutton Publishing Limited 2003

Francis, Paul, Flagg, Richard, Crisp, Graham, *Nine Thousand Miles of Concrete,* Historic England/England Heritage 2016

Gallico, Paul, *The Hurricane Story,* London: Michael Joseph Ltd 1959

Garbett, M and Goulding, B. *Lancaster at War 2,* New York: Charles Scribner's Sons 1980

Gibson, G., *Enemy Coast Ahead,* London: Michael Joseph Ltd 1951

Gould, J., *RAF Bomber Command and its Aircraft 1941–1945,* UK: Ian Allan Publishing 2002

Green, William, *Famous Bombers of the Second World War,* Macdonald & Co. Ltd, London: 1960

Gunston, B., *Fedden,* Derby: Rolls-Royce Heritage Trust 1998

—— *Classic World War II Aircraft Cutaways,* Bounty Books, London: 2011

Halliday, Hugh A., *Woody,* Toronto: CANAV Books 1987

Harris, A. T., *Despatch on War Operations:23 February 1942 to 8 May 1945,* London: Frank Cass & Co. Ltd. 1995

Hastings, Max, *Bomber Command: The Myths and Reality of the Strategic Bombing Offensive 1939–1945* New York: The Dial Press 1979

Hercules XVI, XVII, XVIII, Bristol: manual, The Bristol Aeroplane Co. Ltd. 1944

Higgins, Terry, *The Last of the Buffalo Beaux,* Ottawa: Canadian Aviation Historical Society 2015

Hooker, S., Reed, H., Yarker, A., *The Performance of a Supercharged Engine,* Derby: Rolls-Royce Heritage Trust 1997

Jackson, A. J., *British Civil Aircraft since 1919 Vols 1,2,3,* London: Putnam & Co 1973/1974

Jones, W. E., *Bomber Intelligence,* Earl Shelton: Midland Counties Publishing 1983

Kennedy, Paul, *The Rise and Fall of British Naval Mastery,* New York: Penguin 2004

Bibliography

Killen, John, *A History of the Luftwaffe*, New York: Muller, Blond & White, Ltd 1986

Kostenuk, S. and Griffin, J., *RCAF Squadron Histories and Aircraft 1924–1968*, Toronto: Samuel Stevens/Hakkert 1977

London, Peter, *British Flying Boats*, The History Press, Stroud: 2011

Lumsden, Alec S. C., *British Piston Aero-Engines and their Aircraft*, Shrewsbury: Airlife Publishing Limited 1996

Middlebrook, Martin & Everitt, Chris, *The Bomber Command War Diaries: An Operational Reference Book 1939–1945*, Earl Shilton: Midland Publishing 2000

Mikesh, Robert C., *Excalibur III*, Washington D.C.: Smithsonian Institution Press 1978

Milberry, Larry, *Canada's Air Force at War and Peace*, Toronto: CANAV Books 2000

Motiuk, Laurence, *Thunderbirds at War*, Ottawa: Larmot Associates 1998

Moyes, P. *Bomber Squadrons of the R.A.F. and their Aircraft*, London: Macdonald & Co. (Publishers) Ltd 1964

Murray, Iain R., *Vickers Wellington Owner's Workshop Manual*, Haynes: Sparkford, Yeovil 2012

Neillands, Robin, *The Bomber War*, London: John Murray Pubs Ltd. 2001

Nesbit, Roy Conyers, *RAF an Illustrated History*, Stroud: Sutton Publishing 1998

Overy, Richard, *Bomber Command 1939–1945*, London: Harpercollins Publications Ltd. 1997

Peden, M., *A Thousand shall Fall*, Toronto: Stoddart 1988

Probert, H. *Bomber Harris: His Life and Times*, London: Greenhill Books 2003

Radell, R., *Lancaster, A Bombing Legend*, UK: Chancellor Press 1997

Rapier, Brian J. and Bowyer, Chaz *Halifax & Wellington*, London: The Promotional Reprint Company Ltd 1994

Rawlings, John, *The History of the Royal Air Force*, London: Hamlyn Publishing 1984

Redding, T., *Life and Death in Bomber Command,* UK: Fonthill Media 2013

Richards, Denis, *RAF Bomber Command in World War Two,* London: Penguin Classics 2001

Sharpe, Michael, *History of the RAF,* Bath: Parragon Plus 2002

Stachiw, Andrew L and Tattersall, Andrew, *Handley Page Halifax,* Vanwell Publishing Limited: St-Catherines 2005

Sweetman, J., *Bomber Crew: Taking on the Reich,* London: Abacus 2005

Sweetman, W. *Avro Lancaster,* New York: Zokeisha Publications 1982

Swift, D. *Bomber County: The Lost Airmen of World War Two,* London: Penguin Group 2010

The Air Ministry Account of Bomber's Command Offensive against the Axis, London: His Majesty's Stationery Office 1941

The Greater Vancouver Branch of The Aircrew Association, *Critical Moments,* Vancouver: Self Published 1989

Thompson, J. E., *Bomber Crew,* Canada: Trafford Publishing 2005

Thompson, Scott, *Douglas Havoc and Boston,* Ramsbury: The Crowood Press Ltd 2004

Turner, P St. John, *The Vickers Vimy,* Sparkford: Patrick Stephens Ltd 1969

Wilson, G. A. A., *Lancaster Manual 1943,* Stroud: Amberley Publishing 2013

—— The *Lancaster,* Stroud: Amberley Publishing 2015

—— *Bomber Command: The Men, Machines and Missions 1936–1968,* Stroud: Amberley Publishing 2021

—— *The Merlin,* Stroud: Amberley Publishing 2018

Above: A later Bristol Jupiter engine, cylinders with forged heads and partially enclosed valve gear; a Fedden triumph.
Overleaf, top: 'Clean-looking' 1938 Hercules. *Bottom*: 1945 'Herc' made under licence by SNECMA, mainly for the Noratlas.

INDEX

Aircraft companies, individuals and locations are indexed. It is not possible to index aircraft types within the space constraints.

A.V. Roe and Company (Avro) 16-17, 21, 34, 68, 119-20, 132, 141, 143, 147, 149-52, 187
A.V. Roe Canada Limited 120
Aerospatiale/British Aircraft Corporation 13, 39-41, 122-3, 130, 134-5, 272
Airco 67, 74, 2323
Albert Canal 168
Aldergrove, Northern Ireland 184, 247
Alert, Ellesmere Island, Nunavut, Canada 259
Armstrong, Howie 254
Armstrong, Sir W. G. & Company 117
Armstrong, Sir W. G. Whitworth Aircraft Company 117, 135
Armstrong-Whitworth 117-18, 129, 135, 243
Ashton Down 258
Auckland 195
Austin Motors 143

Barnwell, Flight Lieutenant Richard Antony 35
Barnwell, Frank 25-30, 33, 35, 65, 139
Barnwell, Pilot Officer David Usher DFC 35
Barnwell, Pilot Officer John Sandes 35

Bazalgette, S/L Ian Willoughby VC DFC RAF VR 260-63
Belfast, Northern Ireland 34, 38, 133-4, 191, 193, 197
Berlin Airlift 145, 148, 151, 161, 182-3, 246-7
Birrell, Dave, Archivist and Librarian, BCMC 252, 262-3
Blackburn Aircraft Limited 67-8, 121, 241
Blackburn and General Aircraft Limited 121
Bomber Command Museum of Canada 11-12, 252-273 passim
Boscombe Down 149, 161-2, 164, 173, 177
Botwood, Newfoundland 190
Boulton Paul 139, 144, 175-6, 191
Bowman, Doug, BCMC photographer 252
Breguet Aviation 122-4, 216, 220
Bristol Aeroplane Company 13, 19-41 passim, 49, 52, 59, 116, 122, 154, 158-9, 168, 171, 220, 225, 230, 235, 237-8, 271
Bristol Aeroplane Company Limited 20, 28
Bristol Engines of Canada 54
British Aircraft Corporation 13, 39-41, 122-3, 130, 134-5, 272

British & Colonial Aeroplane Company 13
British and Colonial Aeroplane Company Limited 21-2
British and Colonial Aviation Company Limited 21
Brize Norton, Oxfordshire 247
Broderick (née Bazalgette), Mrs Ethel 260-61
Brough Aerodrome 152
Broughton, North Wales 210
Brunsbuttel, Schleswig-Holstein, Northern Germany 212, 245
Burt, Peter 14, 46, 81, 84
Busteed, Harry 25
Bute, Strathclyde Region, Scotland 241
Butler, Leonard 'Bunny' 10, 45-8, 51, 58, 63-4, 85, 270

Calgary, Alberta, Canada 253, 255, 260
Cambridge 235
Canadian Pacific Air Lines 11, 40, 264
Canadian Pacific Railway Corporation 246, 255
Canary Islands 196
CASA 123-4, 217-8
Chadwick, Roy 119-20, 141-2, 148-9
Churchill, Winston 58, 151, 234
Claresholm, Alberta, Canada 253-4

Index

Clark, Archie 255
Coanda, Henri 25
Collinstown Airport, Co. Dublin, Ireland 13
Cosmos Engineering Company (Cosmos) 14, 16, 27-8, 30, 48-9, 52, 65, 234, 270-71
Cranwell, RAF Cadet College, Lincolnshire 60, 68, 76, 204, 234, 236
Croydon, South London 31, 161

Dartmouth, Nova Scotia, Canada 259
Darwin, Australia 207
de Havilland 67, 126, 140, 154, 206, 211, 237
Douhet, General Giulio 230
Dowding, Sir Hugh 56
Duke of Gloucester 151
Dumbarton, Scotland 241
Dwelle, Harry 257

Edmond, Maurice 22
El Gamil, Egypt 184
English Electric Company 174
Essen, North-Rhine-Westphalia 242, 261

Fairchild Aircraft Ltd (Canada) 34, 128
Fairey Aviation Company 125-6, 165, 174
Farman, Henri 21-2, 233
Fedden, Sir Alfred Hubert Roy 9-11, 14, 16, 18, 27-30, 37, 42-65, 71, 76, 81-6, 234, 237, 270-71, 279
Fiji 195-6
Filton, Bristol 13, 20-28, 31-3, 36, 38-9, 41, 50, 122, 161
Finningley, South Yorkshire 208, 245
Flugzeugbau Nord 130, 222
Fokker 67-8, 127-8, 218, 272
Folland 126-7, 170, 272
Fort Macleod, Alberta, Canada 258, 266
Foster, Ian 265
Foynes, Ireland 190
Frise, Leslie 35, 139

Garratt, Panton 254, 257, 260
General Aircraft 68, 121, 152-3
Geraardsbergen, Belgium 244
Gnome-Rhone company 52, 216, 219-20
Godfrey, Charles 'Chuck' DFC 260
Gotha 124, 232, 241
Greenwood, Nova Scotia, Canada 259

Hafner, Raoul 38
Halton, Cheshire 234
Handley Page Aviation Company 117, 128-30, 172, 174, 272
Handley Page Cricklewood, London 173
Handley Page Limited 16

Harris, Commander-in-Chief, Air Marshal A.T. 142, 242
Hawker 67-8, 118-21, 126-7, 130
Hawkins, Paul 266
Heligoland Bight 242
Hendon, London 46, 235, 262
High River, Alberta, Canada 253, 266

Ismailia, Egypt 207

Jersey, Channel Islands 161
Johannesburg, South Africa 151, 195
Johnson, Herb 256

Karachi, Pakistan 195
Kjarsgaard, Karl 11, 253, 264-6, 268, 273
Knight, Charles 14
Koolhoven, Frederick 68, 80, 117

Lake Mjosa, Norway 252
Lake Naivasha, Kenya 195
Larkhill, Salisbury Plain 22, 25, 230
Le Havre, Normandy 244
Leeming, North Yorkshire 199, 245
Lincolnshire 234, 236, 240, 242, 247
Linton-on-Ouse, Yorkshire 177, 244
Loebelle, Marcel 166

London Aircraft Production Group, Leavesden 174
Longbridge (Birmingham), West Midlands 166
Lord Beaverbrook 57, 155, 246
Lord Cowdray 232
Lord Mountbatten 151
Lord Rothermere 36, 56, 77
Los Angeles 225

Madeira 196
Mahaddie, Hamish DSO DFC AFC and Bar 261
Malton (Toronto), Ontario 150, 257-8
McCollum, Harry 14, 46, 81, 84
McMahon, Ray 256
Metropolitan-Vickers 143
Middleton St George, Durham 177
Morrison, Greg 252

Nanton, Alberta, Canada 11-12, 18, 252-7, 260-62, 264, 266
Napier 14, 127, 168, 170-72
Nash & Thompson 144, 210-11
Nash, Frazer 39, 144
New York 225
Noratlas 17, 112, 114, 116, 130, 219-23, 272, 279
Nord Aviation 130, 219

Index

North Coates, Lincolnshire 240
Northrop 131, 225-7, 272
Northwood, London 239

Oban, Western Scotland 241
Oxford 235

Paris 21, 30, 39, 52, 61, 71, 86, 123, 129, 162, 205, 229, 271
Port Fouad, Egypt 222
Port Said, Egypt 222

RAF Bentley Prior, London 238
RAF Bicester, Oxfordshire 173
RAF Boscombe Down 149, 161-2, 164, 177
RAF Bowmore, Argyll 241
RAF Dishforth, North Yorkshire 247
RAF East Wretham, Norfolk 243
RAF Finningley, South Yorkshire 208, 245
RAF Foulsham, Norfolk 243
RAF Harwell, Berkshire 246
RAF Harwell, Oxfordshire 246
RAF High Wycombe 241
RAF Hurn, Dorset 246
RAF Leeming, North Yorkshire 199, 245
RAF Little Snoring, Norfolk 243

RAF Martlesham Heath 166
RAF Oakington, Cambridgeshire 247
RAF Pembroke Dock, Pembrokeshire 241
RAF Swinderby 204, 247
RAF Syerston, Nottinghamshire 243
RAF Upavon, Wiltshire 245
RAF Upwood, Cambridgeshire 167
RAF Uxbridge 241
RAF Waddington, Lincolnshire 242
RAF Waterbeach, Cambridgeshire 243
RAF Witchford, Cambridgeshire 243
RCAF Station Comox, British Columbia, Canada 259
RCAF Station Scoudouc, New Brunswick, Canada 258
Reid, Wilfred 29-30
Resolute Bay, Cornwallis Island, Nunavut, Canada 259
Ricardo, Harry 14, 82, 84
RNAS Prawle Point, Devon 240
Rochester, Kent 193-4, 197
Rodney Works 34
Rootes 34, 56, 174
Royal Aircraft Establishment, Farnborough 27, 33, 48, 65, 68, 76, 79, 82, 106,

140, 152, 177, 184, 204, 210
Royal Flying Corps 15, 25-6, 119, 134-5, 230-31, 233

SARO 67-8, 73, 132, 188-9
Saunders-Roe 17, 120, 132, 187-8, 272
Schillig Roads 245
Sedan 168
Short Brothers plc 17, 125, 133, 189-90, 192, 194, 197
Shorts 133-4, 151, 188, 190, 194, 197, 198
Singapore 134, 151, 153, 162
Smart, Patrick 266
Smith-Barry, Robert 235
Smuts, General Jan 232
Sopwith 21, 65, 232
South Africa 46, 137, 151-2, 156, 163, 168-9, 180-81, 206, 209, 215, 227, 236
Southampton 126, 132, 195-6
Spokane, Washington, USA 259
St. Athan 258
Stockport, Cheshire 125, 166
Supermarine 67-8, 72, 126, 135, 188, 197, 237, 239
Sydney, Australia 195-6

Terrace, British Columbia, Canada 265, 273
Tinson, Clifford 25
Trenchard, Major General Sir Hugh 232, 234-6
Trenton, Ontario, Canada 12, 252, 258, 264-5

Turbomeca 121, 221, 223-4

Vickers 32, 67-8, 124, 134-5, 170, 175, 178, 180-9, 201, 207, 209-11, 214, 232, 272
Vickers (Aviation) Ltd 17
Vickers Armstrong 39, 117, 143
Vickers Engineering 134
Vickers-Armstrongs Company 39, 117, 135, 201, 203-4, 207, 209
Vickers School of Flying, Brooklands 135
Victory Aircraft 120, 143, 150-51, 258
Vlaardingen, near Rotterdam, Holland 199, 245
Vulcan, Alberta, Canada 253-5, 260

Wallis, Barnes 208, 210-11
Walvis Bay 236
Weston-super-Mare 35, 38, 41
White, George 13, 19-20, 26, 254
White, Samuel 26
Whitehead, Fred 53
Whitley, Coventry, West Midlands 243
Whitworth Gloster Aircraft Company 118
Wilhelmshaven 245
Winnipeg, Manitoba, Canada 162
Wright 54, 64, 199

Also available from Amberley Publishing

Available from all good bookshops or to order direct
Please call **01453-847-800**
www.amberley-books.com